SOFT TISSUE PAIN AND DISABILITY

RENE CAILLIET, M.D.

Chairman and Professor
Department of Rehabilitative Medicine
University of Southern California
School of Medicine
Los Angeles, California

F. A. DAVIS COMPANY Philadelphia

Also by Rene Cailliet:

FOOT AND ANKLE PAIN

HAND PAIN AND IMPAIRMENT

KNEE PAIN AND DISABILITY

LOW BACK PAIN SYNDROME

NECK AND ARM PAIN

SHOULDER PAIN

SCOLIOSIS

Library of Congress Cataloging in Publication
Data

Cailliet, Rene.
  Soft tissue pain and disability.

  Includes bibliographies and index.
  1. Pain.   I. Title.   [DNLM: 1. Connective
tissue.   2. Collagen diseases.   3. Pain.
WD375 C134s]   RB127.C34   616'.047
ISBN 0-8036-1630-9   77-22612

Preface

"The phenomenon of pain belongs to that borderline between the body and the soul about which it is so delightful to speculate from the comfort of an arm chair, but which offers such formidable obstacles to scientific inquiry." So wrote Killgren in the introduction of his classic treatise on deep pain sensibilities.*

A large percentage of complaints that cause a patient to seek medical care is that of pain and disability of a moving part of the body. Medical practitioners of all disciplines glibly discuss pain and impairment on the basis of *soft tissue* injury, stress, sprain, or inflammation, yet soft tissues are not well defined in medical dictionaries, are not named in Nomina Anatomica, and are not so designated in most anatomic or orthopedic texts. Unfortunately the soft tissues are not considered an organ system and, thus, are not on the curriculum of medical schools. Except for occasional symposia or postgraduate courses on musculoskeletal syndromes, the practicing physician receives no education in this area. This results in less knowledge and less interest in the scope of soft tissue damage. Therefore, it is the patient who suffers from this insufficient knowledge and concern—both in pain and in adequate evaluation and treatment.

A basic knowledge of soft tissue and normal functional anatomy of the part(s) involved in pain and disability is mandatory for a meaningful evaluation of the patient. Tissues capable of evoking pain and limitation must be recognized and faulty mechanism of joints must be considered. More meaningful therapy results from this knowledge and many painful, disabling conditions are prevented or decreased in their severity and duration. Many complications requiring more drastic care are prevented. Even psychologic correction can result only if the defective physiologic abnormality is recognized and understood.

The purpose of this text is to evaluate and clarify the basis of many of the painful musculoskeletal and neuromuscular conditions that beset man and perplex his physician. It is hoped that this book, along with being informative, will also stimulate more interest in soft tissue abnormalities.

RENE CAILLIET, M.D.

---

*Killgren, J. H.: *On the distribution of pain arising from deep somatic structures with charts of segmental pain areas.* Clin. Sci. 4:36, 1939.

Table of Contents

v

Illustrations

Introduction

All musculoskeletal pain may be considered a sequelae of soft tissue injury, irritation, or inflammation. Trauma in the broadest concept of the term is the greatest cause of soft tissue pain and functional impairment. Every aspect of the musculoskeletal system is subject to trauma in various forms.

The patient with musculoskeletal pain wanders from physician to physician in quest of relief or, at least, explanation and reassurance regarding his pain and dysfunction. Modalities are employed that have no rationale. Medications are prescribed to decrease pain and spasm while the cause remains unrecognized and, therefore, untreated. Unorthodox practitioners are sought and frequently, much to the dismay of the physician, afford the patient some relief. Nomenclature evolves and becomes ingrained in our culture, often to the disadvantage of the suffering patient. Insurance compensation and litigation compound the abuse of the patient.

In order to better understand the phenomenon of soft tissue pain and disability one must find factors that relate pain and functional impairment in all the joints of the body. These factors develop along this line:

1. Functional anatomy of the involved segment of the body
2. Neuromuscular pattern of the moving parts
3. Tissue sites capable of eliciting pain
4. Responsible faulty neuromuscular mechanism

Each joint in the human body has its own characteristic structure and function and involves soft tissues. These soft tissues include muscles, capsules, ligaments, tendons, menisci, disks, and cartilaginous surfaces. All these soft tissues are subservient to nervous system motor control, innervation for sensation and proprioception, and adequate blood supply. Functional anatomy clarifies the bioengineering principles of the

1

moving parts in their normal movement or posture. The intricacies of movement controlled by the neuromuscular system is precise and allows little deviation. It is controlled by a coordinated central nervous system pattern that is partly developmental; it is modified by training and repetition.

Normal use and normal movement place no stress upon soft tissues. Excessive use, abuse, and misuse can cause irritation with resultant pain and disability. Posture or normal muscular activity may be altered by extraneous factors such as fatigue, distraction, anxiety, impatience, anger, and depression. External stress such as injury, trauma, or mechanical forces irritates the tissues that offer resistance. Once the neuromuscular pattern is altered, the normal biomechanics of the part is altered and the soft tissues are abused and damaged. When these tissues contain pain-mediating nerves, pain results. Appreciation of the tissue sites so supplied with pain nerve endings affords meaningful evaluation of the pain mechanism as well as determination of the specific tissue site.

Therefore it is mandatory that the functional anatomy of the involved structure be known in any musculoskeletal pain. Examination must specifically test the motion of that joint and expectantly reproduce specific symptoms. The history of the expected causative episode must be elicited also. The violation of the neuromuscular pattern that altered the biomechanics of the moving part must be clarified.

Once the abnormal pattern is apparent, treatment must restore the normal muscular pattern by training and repetition until the proper motion becomes automatic. Full range of motion of the impaired part must be restored. Whenever necessary or possible, the damaged tissues must be permitted to recover or be repaired. Many, if not most, musculoskeletal painful disabling abnormalities respond to treatment by mechanical means and surgical intervention is usually not required nor effective.

CHAPTER 1

Soft Tissue Concept

Tissues are the matrix of the body; they are important in determining the disposition and form of all anatomical structures, whether those structures are muscles, nerves, or glands. All tissues are composed of cellular elements and their derivatives. The cells may be closely packed together along their junctional margins by the adhesions of their surface membranes or by direct protoplasmic connections. The cells may be widely scattered through an intercellular ground substance containing tissue fluid, fibrous elements, and organic material. Many tissues are composed predominantly of one type of cell that forms them as a working unit, but mere aggregation of cells does not form a functional tissue. Muscular tissue, for instance, is composed of more than muscle fibers. It also contains intrinsic elements of connective tissue. This is also true of nerve cells and their processes. While it is appropriate to study the physical, chemical, and anatomical properties of isolated cells such as muscle or nerve cells, the basic functional element of muscular and nervous tissues and mechanisms implies organized tissues and takes into consideration other tissues that constitute the normal activity of the living body. A tissue, therefore, may be defined as an assemblage of cellular and fibrous elements in which one particular type of cell or fiber predominates.

The four primary tissues of the body are

1. Epithelial tissue for protection, secretion, and absorption.
2. Muscular tissue for contraction.
3. Nervous tissue for irritability and conductivity.
4. Connective tissue for support, nutrition, and defense.

All of them exhibit structural and functional specialization, yet they are intimately related and interdependent.

CONNECTIVE TISSUE

The largest component of the human body is connective tissue, which also forms a continuum throughout the body. The connective tissue system includes all constituents of the mesenchyme: ground substance, elastin, collagen, muscle, bone, cartilage, and adipose tissue. The various types of connective tissue are not distinct but have many transitional forms. All of them are characterized by large amounts of intercellular material. The consistency of connective tissue depends upon the relative amount and proportion of collagenous (white) and elastic (yellow) fibers. In some sites connective tissue forms a delicate reticulum and in other locations tough fibrous sheets.

The fact that connective tissue must support, nourish, and afford defense against trauma and infection makes it a highly specialized and complex tissue. Connective tissue, therefore, contains as well as comprises blood vessels and lymphatic vessels for its function of nutrition, defense, and repair.

In connective tissue the cells are widely dispersed and separated by a material termed matrix. In the matrix the intercellular substances are of primary importance. The intercellular substance contains fibers and cells embedded in a matrix of semifluid gelatinous substance that is considered by some to be secreted by the cells.

Classification and Function

There are numerous classifications of connective tissue with essentially two types of tissue: fibrous and amorphous. The proportion of fibrous to amorphous tissue at a particular site depends on the role and function required of that tissue.

Collagen was defined by Virchow in the nineteenth century as "body excelsior" or "inert stuffing." Collagen comprises fibrils of formed elements and amorphous substance. Ground substance is in turn divided into interfibrillar substance termed "cement" and the ground substance of the fibers. Ground substance contains polysaccharides of which acid mucopolysaccharide is the best understood. Recently there have been major advances in understanding the structure, biosynthesis, and metabolism of ground substance. It is now recognized as a family of proteins with tissue specificity.

Connective tissue may be grouped as follows:

I. Connective tissue proper
   A. Loose connective tissue (areolar): contains many spaces that may contain fluid and is involved in the cellular metabolism
      1. Intercellular substances are

      a. Collagenous or white fibers: collagenous are primarily parallel fibers bound together in bundles giving it great tensile strength

      b. Elastic or yellow fibers: elastic yellow contributes to its elasticity

      c. Reticular fibers: essentially delicate collagenic fibers and function to support cells

B. Dense connective tissues

C. Regular connective tissue

    1. Tendon (white fibrous connective tissue)

    2. Fibrous membranes

    3. Lamellated connective tissue

II. Special connective tissue

  A. Mucous

  B. Elastic: fibers run singly, branch freely, and anastomose with each other; under ultramicroscope looks like a twisted rope

  C. Reticular

  D. Adipose

  E. Pigmented

III. Amorphous

  A. Ground ⎫

  B. Cement ⎭ both are gels of varying concentration and rigidity

IV. Cartilage

V. Bone

VI. Blood and lymph

Connective tissue has various functions:

1. Provides supporting matrix for more highly specialized organs or structures
2. Provides pathways for nerves, blood vessels, and lymphatic vessels by its fascial planes
3. Facilitates movement between adjacent structures when it has a loose structure
4. Forms bursal sacs that minimize local effects of pressure and friction
5. Creates restraining mechanisms of moving parts by the formation of bands, pulleys, and check ligaments
6. Aids in promoting the circulation of veins and lymphatics by ensheathing the limbs
7. Furnishes sites for attachment of muscles
8. Forms space for storage of fat to conserve body heat
9. Has fibroblastic activity and thus repairs tissue injury by forming scar tissue

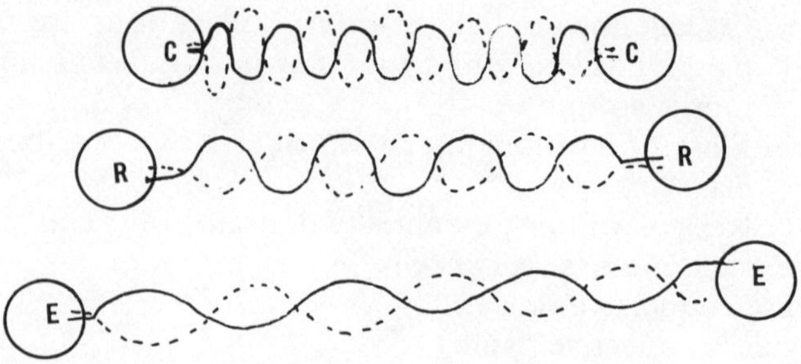

FIGURE 1. Connective tissue fibers which elongate by straightening. C, The shortened connective tissue fiber; R, The resting length; E, The elongated fiber.

10. Contains histiocytes (a connective tissue cell) which participate in phagocytic activity in defense against bacterial invasion
11. Synthesizes antibodies to neutralize antigens by its plasma cells (another connective tissue cell)
12. Contains tissue fluids that participate in the nutrition of tissue

In fibrous connective tissue the spaces between the cells are occupied by numerous fibers that allow the tissues to withstand distortion and strains. Fibers vary from loosely woven to closely packed with little interfibrillar space (Figure 1). As connective tissue ages its fibers shorten, become more densely packed, and thus interfere with movements of associated musculoskeletal parts. Fibrous connective tissue, being ubiquitous, is almost always involved in injury and plays an important part in the process of healing. Specialized tissues of the body or its organs that cannot be replaced by specialized tissues become repaired or replaced by connective tissue called scar.

Varying Types of Connective Tissue (Fig. 2)

FASCIA. Specialized tissues are enveloped in connective tissue sheaths termed fascia. Individual muscle groups can be enveloped in fascia that separates that group from an adjacent muscle group. There is fluid between the fibers of this fascia that acts as a lubricant to permit freedom of movement of adjacent muscle groups (Fig. 3). From the outer fascial connective tissue septa pass down into the muscle and progressively divide it until the septa become very delicate and ramify to surround individual muscle fibers.

TENDONS. Tendons are bundles of heavy collagen fibers that run parallel to one another. Between the bundles are fibroblasts. Tendons principally function to connect muscles to bones and to concentrate the pull of a muscle upon a small area. Tendons can sustain 8700 to 18,000

6

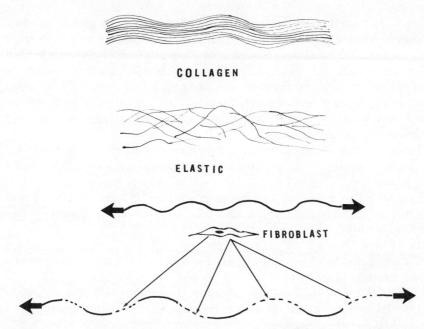

FIGURE 2. Varying types of connective tissue.

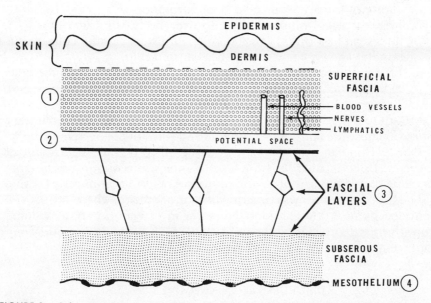

FIGURE 3. Schematic layers of fascia. *1*, Superficial layer containing fat, pressure nerve receptors, and blood vessels; *2*, "Potential" layer may become large space when dissected by extravasation or edema; *3*, Deep investing layer; and *4*, Layers of pleura, peritoneum, pericardium, and so forth.

pounds per inch so they are rarely ruptured by excessive tension. Clinically when a tendon tears its tearing occurs from the muscle or from its bony attachment. There is a limit to the resistance fibrous tissues can offer to tensile forces. Ligaments and fibrous bands exposed to continuous excessive (abnormal) tension stretch fairly rapidly. This is not due to elongation but due to proliferative fibroblastic activity of fibroblasts, which produce more collagenous tissue and thus increase the length of the entire structure.

Tendons assume their glistening appearance because they are invested in a loosely textured connective tissue that contains a glairy fluid similar to synovial fluid.

LIGAMENTS. Ligaments are similar to tendons, but the collagen fibers are not so regularly arranged and may contain some elastic fibers. Ligaments usually connect bones to other bones.

The fibrous capsule of joints are composed mainly of collagen fibers that run from one bone to another. They have local thickening which forms intrinsic ligaments.

CARTILAGE. Cartilage is a fibrous connective tissue with abundant firm matrix. Its cells are called chondrocytes and they lie within spaces called lacunae. Types of cartilage are

1. Hyaline, which has a homogeneous matrix and considerable elasticity. Cartilage precedes bone formation.
2. Fibrocartilage, which is limited in the body, is located mostly in intervertebral disks, and has thick compact parallel collagenous bundles and very few, if any, encapsulated cells.
3. Elastic, the smooth and glistening cartilage which covers the ends of bones and thus becomes a component part of joints.

MUSCLE: There are two types of muscles: smooth and voluntary striated. These are considered a specialized form of connective tissue.

BONE: Bone and cartilage are considered to be modified forms of collagen. Bone is a harder bundle of connective tissue in which large amounts of calcium comprise a solid matrix of fibrous connective tissue.

The fibroblast which was traditionally considered the characteristic cell of mesenchyme is now considered by some investigators to be a modified smooth muscle cell. It is apparent that much remains to be learned concerning connective tissue structure, origin, and metabolism.

DISORDERS OF CONNECTIVE TISSUE

Circumscribed or systemic lesions of connective tissue give rise to a large number of diseases, collectively summarized in a concept of connective tissue disorders. Otaka classified the abnormalities by connective tissue in the following order[1]:

1. Metabolic disorders
2. Aging of connective tissue
3. Hereditary disorders
4. Systemic inflammatory disease
5. Organ fibrosis
6. Tumors of connective tissue
7. Miscellaneous connective tissue diseases

At the present time systematization and classification of all connective tissue changes and disease is not complete.

There are inherent disorders of connective tissue which also have a metabolic basis, but these comprise a small percentage as compared to the vast array of acquired connective tissue impairments that present as clinical entities.

Generally the primary site of pathologic processes has been considered to be the ground substance, especially the interfibrillar matrix, but fibrous elements of tissue also contribute to pain and functional disability. The symptomatic connective tissue disorders that result from trauma, misuse, or abuse are less well understood and documented. All these are grouped together as soft tissue injuries or diseases causing musculoskeletal or neuromuscular pain and dysfunction.

Body Mechanics

Moving parts of the body which bear the brunt of connective tissue disability must be studied carefully. All aspects of the kinetic musculoskeletal system are involved.

Kinesiology is the scientific study of movements. This concept includes familiarity with skeletal anatomy and the neuromusculoskeletal system. Neurophysiology, physiology, chemistry, and the sciences of physics and mechanics relate to this neuromuscular system function of movement.

Movement of the body occurs at joints and depends on muscular action. The muscular action in turn is dependent on the peripheral and central nervous system. This complex body mechanism must be considered in the concept of (1) normal activity and (2) how deviation from normal results in pain, impairment, and disability.

Joint Motion

The type and extent of normal movement occurring in any joint depends upon (1) the form of the articular surfaces, (2) the restraining influences of the ligaments, (3) the atmospheric pressure within the joint, and (4) the control exerted by the muscles as they act upon the joint.

The original division of the forms of surfaces that articulate to form a joint were the following:

1. Plane joints, in which the articular surfaces are flat and permit very limited motion
2. Spheroid joints, the so-called ball-and-socket joint, in which a well-rounded convex surface articulates with a concave surface
3. Cotylic joints, a spheroid joint in which the articular surfaces are relatively longer in one direction
4. Hinge joints, in which movement usually is in one plane but with some slight rotation
5. Condylar joints, a hinge type with two distinct articular surfaces
6. Trochoid joints, in which there is rotatory movement of a ring-shaped structure around a bony pivot
7. Sellar joint, a joint resembling horse saddle-shaped surfaces

Currently joint motion has been reduced to only two basic types: ovoid and sellar (Fig. 4).

As well as influencing the direction and extent of joint motion, the congruity of joints also has significance in articular lubrication. Two kinds of displacement are permitted by the shape of the articulating surfaces: spinning and sliding (translation) (Fig. 5).

The joint restraint imposed by ligaments varies in the numerous types of joints throughout the body. Ligaments, once thought to be composed of series of parallel fibers, are now known to have a spiral arrangement of fibers. Rotation of one bone upon another, therefore, has profound effect upon the laxity or tautness of the ligaments (Fig. 6).

Normal joint range of motion combines physiological flexion-extension with simultaneous rotation. However, excessive motion in

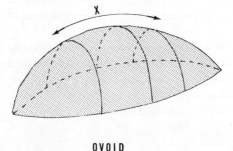

OVOID

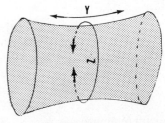

SELLAR

FIGURE 4. Joint motion is considered to be of two basic types: ovoid and sellar. Ovoid permits motion in one plane, X; sellar permits motion in two planes, Y and Z.

10

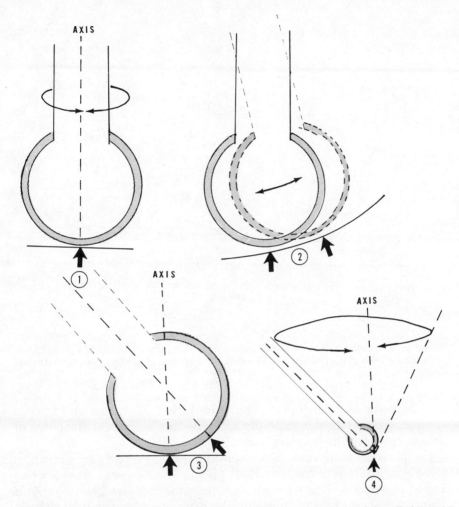

FIGURE 5. Two kinds of displacement: *1*, True spin about an axis; *2*, Arc slide called "swing" when there is no simultaneous rotation or spin; *3*, Spin about the axis and rotation; *4*, Spin is rotation about an axis perpendicular to fixed surface. (Adapted from Licht, S. (ed.): Arthritis and Physical Medicine. Williams & Wilkins, Co., Baltimore, 1969).

either flexion-extension or abnormal rotation during flexion-extension can damage the ligaments.

Ligamentous attachments to bones are generally strong and injury often leads to bone fracture rather than rupture of the ligament. Tendons are noncontractile tissues so their intrinsic stretch tolerance is important. Ligaments are strongest in their middle and weakest at their ends, the site of insertions. Fascial and ligamentous tissues undergo inflammatory shrinkage. Ligaments and fibrous capsules, once damaged, are slow to

11

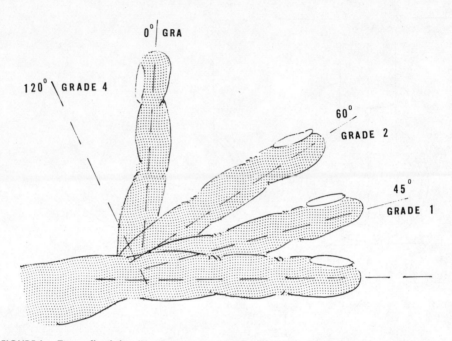

FIGURE 6. Tissue flexibility (Harrington criteria). The flexibility of soft tissue is graded 1 to 4 by hyperextension of the fingers. This is an arbitrary method of grading generalized tissue elasticity.

heal because of poor blood supply. However, ligaments and fibrous capsules have a rich nerve supply and, thus, are an important source of pain so characteristic of many joint diseases.

Constant pressure or constant tension upon ligaments are known to cause wasting, whereas intermittent pressure and tension cause growth and an increase in their strength, especially in their attachment to bone. This fact plays an important role in the reaction of ligaments to stresses and injuries and in their response to treatment.

MUSCULAR ACTION. Muscle *contraction* is a physiologic voluntary reversible change in muscle length from the normal resting state. Muscle *contracture* is an abnormal permanent state of muscle length that deviates from normal. Contracture may be caused by:

1. dynamic imbalance: failure of the agonist-antagonist relationship
2. impaired innervation: peripheral or central
3. intrinsic changes with fluid or cellular infiltration and fibrosis

Muscles by their origin and insertion and their alignment across joints move those joints. They also insert into the fibrous capsules of the joint and protect the joints by this insertion. When a capsule is stretched

12

excessively the muscles attached to them contract as a result of the stretch and, thus, make the capsule more taut. These muscles are termed "shunt" muscles. They act along the lines of moving bones. In rapidly moving joints they neutralize the forces that tend to draw the articular surfaces apart. The effect of atmospheric pressure, which keeps the surfaces approximated, is thus protected. In joints whose stability relies exclusively upon muscular control, such as the shoulder glenohumeral joint, dislocation occurs with minimal ligamentous injury. In joints largely dependent upon ligaments for stability, dislocation results in tearing of ligaments with a poor prognosis for recovery or repair.

Learned Activity. Neuromusculoskeletal function requires normal joint mechanics throughout the body and well-coordinated neuromuscular activities. These neuromuscular activities require good reflex activities of a genetic nature, which have been coordinated and modified by properly learned patterns of activity.

Man is what he is because of his brain. A study of the evolution of merely the postural aspect of man gives a clue not only to the evolution of man but also to all primates. Although speech has been considered the primary difference of man from all other animals, it is probably more correct to state that it is the nervous system of man that makes the difference.

The central nervous system is influenced by numerous stimuli, but the bulk of stimuli affecting the function of the central nervous system is from muscular activity that is constantly affected by gravity. Muscular control develops slower in man than in most other animals. The human brain at birth is approximately one fifth of its ultimate weight; however, an animal is born with a more fully grown brain, which soon after birth is ready to react to external stimuli. The animal's behavior is reactive and is a physiologic process that reacts to the environment. Such animals have little capacity for learning. Learning, as a general concept, is the acquisition of new responses to stimuli; therefore, learning is also possible only by extinguishing or diminishing basic instinctive responses. An animal brain at birth can do all it will do as an adult, with growth essentially improving only rapidity, precision, and reliability of neuromuscular functions. Man's brain develops in many other ways.

Man modifies and controls his instinctive actions and reactions. His central nervous system growth is influenced by external stimuli. Environmental and individual experiences play a large part in the development of function of man. Muscular activity has little, if any, genetic inheritance but develops as a learned activity.

Learned activities can be considered as skills. It is accepted that, in the process of learning a skill, almost all behavior is motor in nature (Fig. 7). Man responds with voluntary and involuntary movements that includes basic posture.

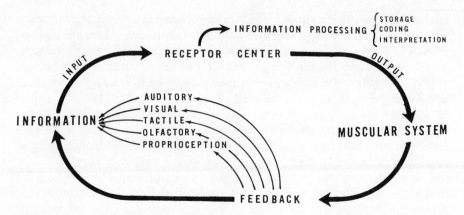

FIGURE 7. Information from internal and external environment via receptors. Process of learning a skill. Almost all behavior is motor in nature; man responds with voluntary and involuntary movements, which includes posture among others. Learning skills proceed in phases (see text). Feedback is one of the most important concepts in learning and is an important factor in the control of movement and behavior.

Learning skills proceed in three phases:

Phase One: Recognition of the plan with its objective and goal. The efficacy of the information source must be determined. Sensory input has priorities since a man cannot contend with numerous stimuli simultaneously. He selects one stimulus, most frequently a visual stimulus, and excludes or dampens the others. The attention span also influences how effective is the one source of information.

Phase Two: Practice. This is influenced by the complexity of the act, the desire or motivation to act, and the importance of the action. Feedback is important in the practice phase in determining frequency and intensity as well as correctness of the practiced act.

Phase Three: Execution. In this phase the neuromuscular system's output increases in ease and efficiency and decreases stress and anxiety. Actions become increasingly more automatic. If the action is proper, the proprioceptive phase is so coded. If the action is improper, an improper coding results and subsequent action, although incorrect, is interpreted as correct by the individual.

Feedback is one of the most important concepts in learning and is an important factor in the control of movement and behavior. This ability to form individual nervous pathways and patterns make faulty function possible as a learned process which persists throughout life. The "feeling" that a person has of an action or function, such as posture, may "feel normal" but may be abnormal, and becomes a deeply ingrained pattern.

14

The impression that bombards the central nervous system is a feedback mechanism coming from the exterior, but it may emit also from internal sources.

These feedback impulses may be subconscious and make activities automatic or reflex. They may evoke a sensory pattern of normal function, abnormal function, or even painful function.

Neuromusculoskeletal pain can adversely affect the psyche or the reverse can occur: the psyche can influence the neuromuscular pattern resulting in pain and disability. Both aspects of the cycle must be understood and appreciated in the treatment of painful states.

The current emphasis on "body language" is an indication of the recent awareness that the body portrays the emotions. Anxiety and its concomitant musculoskeletal tension play a large role in painful states that frequently elude the examiner or are given inappropriate credence and significance. How this portrayal causes painful body stance or faulty neuromuscular activities is not always fully understood nor appreciated.

Neuromuscular actions that are learned or acquired and, thus, become automatic may be disrupted or impaired by anxiety, distraction, or depression. As a result, the usual automatic neuromuscular activities may become faulty, irregular, and erratic and cause musculoskeletal malfunction with resultant pain.

All patterns of impulses that reach the central nervous system from the muscles, joints, and viscera are associated with an emotional reaction. Voluntary skeletal muscular contraction can exert control over the neuromuscular actions caused by the emotions. This is the basis of many forms of exercises advocated to relieve states of emotional tension. Unfortunately, however, many neuromuscular patterns that result from an emotional basis are not amenable to voluntary muscular effort in our present state of knowledge.

Many painful disabling conditions of man are the result of trauma. Misuse or abuse of connective tissues are not yet fully recognized nor appreciated by most physicians. The attempt is made to relate many painful conditions to other organ systems or to relate pain to psychologic interpretation with exclusion of musculoskeletal relationship or origin when, in reality, the musculoskeletal component is an important aspect of the pain and disability.

POSTURE. Painful disabling conditions of the soft tissues of the musculoskeletal system relate in a direct or indirect way to the upright posture of man in standing, moving, and sitting. Posture in man is unique among animals and locomotion is precarious. Erect stance, even though it is unstable, is assumed to have been man's adaption for free use of his arms.

Posture—the way we sit, kneel, squat, and stand—is determined not only by the human anatomic structures but also by culture. People of the world differ in posture style as they do in clothing, housing, occupations, and musical preference.

The ordinary upright posture with arms hanging loosely at the side or clasped in front or behind is universal. Sitting in a chair is not. One quarter of the human race habitually takes weight off its feet by crouching in a deep squat at rest or at work (Fig. 8). Chairs, stools, and benches were in use in Egypt and Mesopotamia 5000 years ago, but the Chinese used chairs only as recently as 2000 years ago. Before that time they sat on the ground as do the Japanese and Koreans. The cultures of the Middle East, North Africa, and Islam have returned to sitting on the floor "for cultural prestige."

Work and rest in a deep squat position is used by millions of people in Asia, Africa, and Latin America. The Turkish or "tailor" cross-legged squat is used in Asia, Korea, Japan, and the Middle East to India. Legs crossed or folded to one side, which was thought to be assumed by women because of narrow skirts, is used by cultures who wear no clothing.

Standing is influenced by the use of footwear as well as a complex of many factors: anatomic, physiologic, cultural, environmental, occupational, technologic, and sexual.

Of interest is the cultural attitude towards posture. Religious concepts

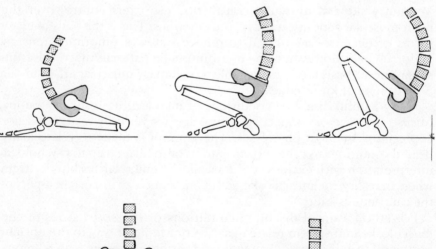

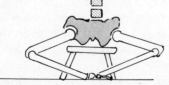
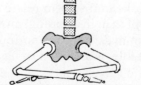

FIGURE 8. Sitting postures of oriental cultures.

influence posture such as kneeling, bowing, standing, and positions used in prayer. Recently Western culture has relaxed postural codes. Eighteenth century chairs with hard seats and straight backs have been replaced by soft curved chairs or sofas. We still, however, train our children to conform to cultural norms of posture by verbal instruction.

Standard posture is one of skeletal alignment refined as a relative arrangement of the parts of the body in a state of balance that protects the supporting structures of the body against injury or progressive deformity. This was the definition given by the Posture Committee of the American Academy of Orthopaedic Surgery, 1947.

The morphologic relationship of the parts of the body to each other are related to gravitational stresses. The center of gravity of the human body is shown in Figure 9.

The erect posture is aimed at musculoskeletal efficiency. In stance the erect posture is maintained largely by ligamentous support.

Lumbar lordosis is maintained by "resting" on the anterior longitudinal ligament with slight weight leaning upon the posterior articulations (facets). The hips are extended to bear weight on the iliopectineal ligament (a thickening of the anterior capsule of the hip joint—the Y ligament of Bigelow). The knee is locked in extension upon the posterior popliteal ligaments and capsule. Only the ankle joint cannot be locked, but this joint can be maintained by slight intermittent isometric contraction of the gastrocnemius and soleus muscles posteriorly and the anticus anteriorly. As the leg shifts backwards the anterior levator pulls the leg forward.

As the ligaments alternately bear the brunt of support at the lumbar spine, the hip, or the knee, the feedback to the brain causes the posture to shift from ligamentous support to muscular support. This implies decrease of lumbar lordosis and flexion of hip and knee. These degrees of flexion are slight and immediately upon muscular fatigue revert back to ligamentous support.

The superincumbent curves—cervical and lumbar lordosis and dorsosacral kyphosis—are seen in the evolution of the child from birth to ultimate erect posture.

The body is poorly engineered for standing because stance is maintained with the heavy parts at the top upon a narrow base (Fig. 10).

Balance is more efficient with less energy expenditure if none of the parts are too far from the vertical axis. The muscular contraction needed to align the head vertically above the trunk and pelvis is minimal if all the parts of the body are close to the vertical center of gravity.

Increase in thoracic kyphosis decreases pulmonary rib cage excursion but also alters the motion of the shoulder girdle; the round-shoulder posture alters the glenohumeral joint mechanism by anteriorly depressing the overhanging acromion and simultaneously internally

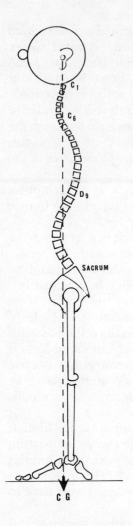

FIGURE 9. Erect posture through the center of gravity. It passes through the external meatus, through the odontoid process, slightly posterior to the center of the hip joint, slightly anterior to the center of the knee, and slightly anterior to the lateral malleoli.

rotating the dependent arm. Both of these postures contribute to cuff entrapment and attrition. This is discussed in further detail in Chapter 6.

The cervical spine is as dependent upon posture as is the lumbar spine. Increased cervical lordosis decreases the intervertebral foraminal openings and predisposes for nerve root radiculitis and further degenerative changes. This is discussed further in Chapter 4.

The posture of man as he stands, walks, sits, works, and rides is important to understand because (1) it affects function of the extremities and (2) it portrays the emotions, which beneficially or adversely influence the learned neuromuscular mechanisms that motorize the musculo-skeletal system.

The deep-seated pattern of emotionally evoked postures has been

18

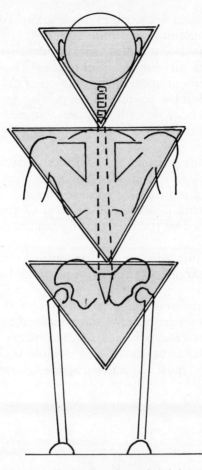

FIGURE 10. Body engineering of stance performed with broad heavier parts at the top situated upon a narrow base.

expounded by Feldenkrais.[2] He states that improper head balance is rare in young children except in structural abnormalities. However, repeated emotional upheavals cause the child to adopt attitudes that ensure safety. This, he claims, evokes contraction of the flexor muscles inhibiting extensor tone. His analogy to animals is that when they are frightened they react by violent contraction of all flexor muscles, thus preventing (inhibiting) the extensor musculature. This prevents running or walking. A similar reaction occurs in newborns as a reaction to the fear of falling.

The attitude of the child from repeated emotional stresses is that of flexion with concurrent inhibition of the extensors. This attitude in the upright erect posture becomes one of flexion at the hips and spine with a forward head posture. This posture becomes habitual and feels "normal."

Since the antigravity muscles work in this imbalanced posture without respite, fatigue and muscular discomfort results.

During sitting the body weight is supported upon the ischial tuberosities and their surrounding soft tissues. The superincumbent lumbar curve is dependent upon the angulation of the sacrum which, in turn, is dependent upon the posture of the pelvis.

Sitting posture has been described as anterior, middle, and posterior, depending on the relationship of the center of gravity to the ischia (Fig. 11).

In the anterior sitting posture, the center of gravity is anterior to the ischia, the lumbar lordosis is decreased, and more than 25 percent of the body weight is transmitted via the feet to the floor.

In the middle sitting posture, 25 percent of the body weight is transmitted to the floor by the feet and the lumbar curve is straight or slightly kyphotic.

In the posterior sitting posture, lumbar lordosis is definitely reversed and less than 25 percent of body weight is transmitted to the floor via the feet.

Intradiskal pressures and muscular activity of the erector spinae muscles, as related to sitting posture, have revealed interesting findings. Intradiskal pressure is greater in any sitting posture than in standing. Merely supporting the arms on the thighs while sitting decreases the intradiskal pressure and muscular activity, as does assuming more lordosis from sitting erect.

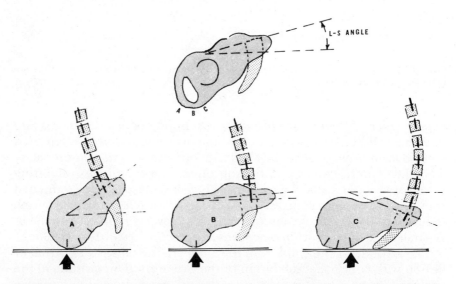

FIGURE 11. Sitting posture is dependent upon the relationship of the center of gravity to the ischia (below, arrows). Sitting postures may be A, anterior; B, middle; and C, posterior.

20

With back support, less muscular activity is required to sit and there is less intradiskal pressure. Inclination of the back of the chair is also related to energy efficiency (Fig. 12). There is less muscular activity required as the chair back angles from 90 to 100 degrees; after 100 degrees reclining there is no significant benefit.

In lumbar kyphosis, intradiskal pressure increases while pressure decreases as the spine moves toward lordosis. The effect of lumbar hyperextension is not known.

Walking (horizontal translation) is most efficient when the center of gravity moves normally in vertical direction. The determinants of gait are credited with accomplishing the minimizing of vertical ascent and descent of the body during gait.

Less energy is required for initiation of horizontal translation if there is a balance of all elements close to the vertical center of gravity. More energy is required to initiate rotation when there is the least "movement

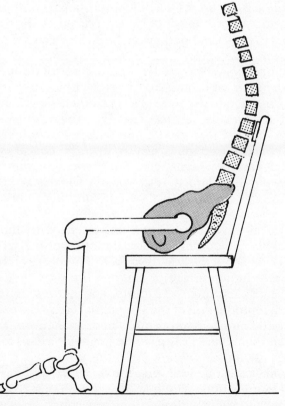

FIGURE 12.   Proper sitting posture with slight lumbar lordosis. Contact with back of chair to upper lumbar region, and 10 degree incline of the chair back; thighs horizontal with lower leg at 90 degree angle.

of inertia," i.e., movement is closer to the vertical axis. Once movement is initiated, less energy is required to maintain forward translation or rotation.

METHOD OF EXAMINATION

The standard method of patient examination includes a history, physical examination, and laboratory evaluation. This general format is also utilized in establishing the diagnosis of musculoskeletal pain and disability. But the examination of the musculoskeletal system must be far more inclusive.

The history should tell the examiner the details of how the initial occurrence of pain was noted and what exact posture, movement, and/or activity initially caused the symptoms and now aggravate it or maintain its presence. The circumstances that existed and possibly could have influenced the initial incidence can be noted. Those circumstances such as the incidence occurring during severe anxiety, depression, haste, distraction, and so forth may not be appreciated by the patient nor even by the unwary physician, but they can be elicited and definitely related to the incidence.

A careful history is always to be expected from a competent physician, but the examiner must be aware of the significance of these factors or his questioning may elude the stated facts and their significance. Neuromuscular functions that have become stereotyped by repetitive use may be, at the time of "injury," distorted by extraneous stimuli such as a recent argument, concern over one's health, financial concerns, anxieties, or distracting factors. The statement, "But I have done this type of activity for twelve years" should not detract from the fact that at the time of injury some other extraneous activities were influencing an otherwise routine activity.

The physical examination of the patient with musculoskeletal symptoms is also unique in that actual motion of the specific anatomic section is performed actively and passively. This phase of the examination is to evaluate the adequacy of the moving part and, thus, test the ligaments, capsules, muscles, joints, and nerves. Also, during this examination, reproduction of the symptoms reveals the exact mechanism of pain producing the symptoms. These maneuvers are specifically diagnostic and simultaneously give the examiner a basis for a meaningful therapeutic program.

The examiner must be well versed in *functional anatomy* and skilled in meaningful examination techniques that test the soft tissues of the body. Unfortunately, today's medical school curriculum does not sufficiently stress functional anatomy.

The third phase of the examination, that of laboratory verification, involves roentgenographic studies and laboratory tests of blood elements, metabolic factors, and so forth. Unfortunately, the usual x-ray pictures do not reveal soft tissue changes, except by exclusion. Bony structures are revealed and related soft tissues must be diagnosed by interpretation. Also, regrettably, many terms employed by the radiological report are incorrectly interpreted and given to the

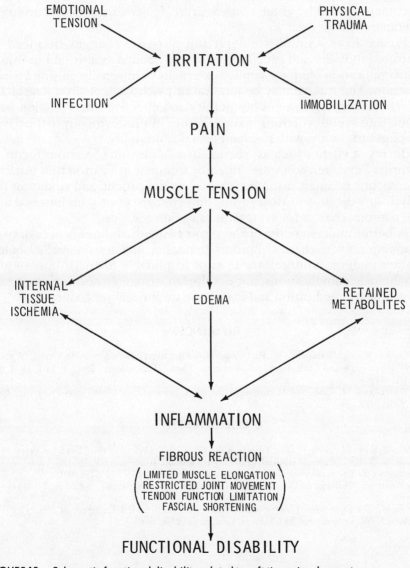

FIGURE 13.   Schematic functional disability related to soft tissue involvement.

patient as "the diagnosis and basis of symptoms" in an attempt to reassure the patient that his "medical condition is clearly understood" and that the treatment is specific. Neither of these deductions is accurate nor true. The term "arthritis" of the spine given a 48-year-old patient to explain low back pain is not accurate, meaningful, reassuring to the patient, nor the true basis of specific treatment.

Musculoskeletal and neuromuscular dysfunction impairs man as much as organ diseases of infectious or metabolic etiologic factors, yet it becomes clouded by vague nomenclature, faulty concepts, and arbitrary therapies.

Figure 13 is a simplistic depiction of tissue changes that lead to impaired mobility and function. *Trauma* is defined as a wound or injury with implication of a force applied externally or internally causing a tissue reaction. The trauma may be physical or psychologic. Pain is a resultant which has varying degrees of intensity and effective interpretation with numerous avenues of transmission. Figure 13 alludes mostly to the tissue response to trauma with resultant pain and disability.

Terms or cliches such as "rheumatism," "fibrositis," various forms of "arthritis," and "tension state" all enjoy frequent utilization that justify a therapeutic regimen, assumingly reassures the patient, and vindicates the physician's diagnosis. Most terms, however, are poorly understood and the pathomechanics for symptoms causation not clear.

A better understanding of these soft tissue disabilities is necessary so that patients with these symptoms do not become detrimentally labeled, do not undergo unnecessary tests of formidable financial burden and psychologic complications, nor develop an iatrogenic disorder as a result of ill-advised medication and inappropriate surgical procedures.

REFERENCES

1. Otaka, Y., and Watanabe, Y.: Pathology of connective tissue disease. In Otaka, Y. (ed.): Biochemistry and Pathology of Connective Tissue. Igaku Shoin Ltd., Tokyo, 1974. pp. 152–179.
2. Feldenkrais, M.: Body and Mature Behavior, ed. 3. International Universities Press, Inc., New York, 1975.

BIBLIOGRAPHY

Clark, W. E., and LeGros, E.: The Tissues of the Body, ed. 6. Oxford of the Clarendon Press, 1971.
Cronkite, A. E.: The tensile strength of the human tendon. Anat. Rec. 64:173–186, 1936.
Klemperer, P.: The concept of collagen disease. Am. J. Path. 26(4): 505–519, 1950.
Klemperer, P., Pollack, A. D., and Baehr, G.: Diffuse collagen disease. J.A.M.A. 119:331–332, 1942.
Robb, M.: The Dynamics of Motor-Skill Acquisition. Prentice-Hall, Englewood Cliffs, NJ, 1972.
Weiner, N. W.: Cybernetics. M.I.T. Press, Cambridge, MA, 1948–1961.

CHAPTER 2

Chronic Pain Concept

Chronic pain is the most serious disabling disease of humans. Pain can no longer be considered merely a symptom; it may actually be a disease itself. Intractable pain may lead to narcotic addiction, expensive disability, or ill-advised surgery of questionable value.

How one thinks about pain depends on the evaluator's learning, experience, or specialty. If the evaluator is a neurologist or neurosurgeon, pain is a result of a neurophysiologic abnormality. To a psychiatrist or psychologist, pain is an emotional affect resulting from internal emotional conflicts. The behaviorist may consider pain as a manipulative phenomenon just as the organically-oriented physician will consider pain as an organ-language to avoid death and impairment.

Steinbach[1] classified *pain* as an abstract concept which the patient describes as a personal sensation of "hurt" that may signal tissue damage with the aim to protect the organism from harm.

The interpretation of pain by the patient and by the observer as well as the methods of treatment remain as varied as are the concepts of pain. Regardless of the neurophysiologic, physiologic, behavioral, or psychiatric basis of pain, noxious irritation of tissues plays a major role in many painful conditions that initiate the remaining mechanism.

The physiologic mechanism of pain is undergoing a serious revision. The previously accepted concept that the impaired tissues affect special nociceptors in the skin, there to be transmitted within the cord to a specific pain center in the brain, is no longer accepted in its simplicity.

CLASSIFICATION AND THEORIES OF PAIN

Bonica[2] classifies chronic pain into three groups:

1. persistent peripheral noxious stimulants
2. neuraxis pain
3. learned pain behavior

In the first group he lists long-term medical conditions such as arthritis, herniated disk, and cancer. In the second group there is involvement of the nervous system: the peripheral nerves, cord, or the brain. The third group he classifies as the patients who receive reward for their being sick or impaired. All three may play a significant role in most painful states but do not necessarily complete the possibilities of causative explanation, nor are any of the three fully delineated.

There are two major neurophysiologic theories of pain transmission:

1. The *specificity theory* claims that under each sensitive spot there lies a unique nerve terminal which, when stimulated, arouses a specific sensation. This was classified to include touch-sensitive Meissner's corpuscles, cold-sensitive Krause's end bulbs, and heat-sensitive Ruffini's end organs. These end organs have never been histologically or physiologically confirmed nor found to be of a specific modality. More probably, the peripheral nerve endings that transmit sensation are chemioreceptors irritated by variable intensity of stimulae from irritation of tissue within which they reside.

2. The *pattern theory* implies a specific sensation, such as pain, as an encoded spatiotemporal transmission. Once a pattern is "set" any stimulation may be similarly interpreted, although difficult to substantiate objectively. There is a relationship between the intensity of the stimulus and the magnitude of sensations. This concept currently is the only application of the specificity theory.

There are two major neurophysiologic transmission systems (Fig. 14):

1. Spinothalamocortic transmission system—involves few neurons, has a short latency, and precisely discriminates intensity and duration.
2. Spinoreticulothalamic transmission system—has complex multisynaptic connections, does not precisely discriminate location, and has a long latency. It projects to the limbic system and is involved in the emotional system.

There are several accepted classifications of nervous fibers:

Fiber A: myelinated afferent, efferent, somatic
Fiber B: myelinated preganglionic and thus sympathetic
Fiber C: unmyelinated somatic, afferent, and/or sympathetic

Fibers A and C transmit pain sensation. Fibers A are further divided into

alpha (12–21 μ) in diameter
beta (8–12 μ)
gamma)   5–10 μ transmit impulses rapidly and localize sensation
    delta)

26

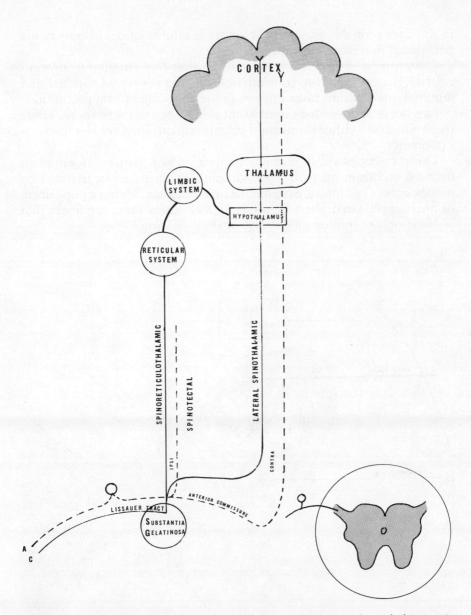

FIGURE 14. Two major neurophysiologic transmission systems. A, Spinothalamocortic system, has spatiotemporal localization; C, Spinoreticulothalamic system, has no localization but is involved in emotional (limbic) and avoidance reaction.

Since C fibers are found in the fasciculus proprius on *both sides* and are considered to carry pain sensation, it is possible that pain is transmitted up both sides of the cord. This multiplicity and bilaterality of pain tracts in the cord probably explains the unpredictability of cordotomy being performed to treat chronic pain.

Currently, it is reasonable to assume that fine cutaneous afferent activity is necessary for transmission of pain sensation. Spatial and temporal mechanisms must operate in the encoding of pain perception.

Free nerve endings that can transmit pain are found everywhere. These free endings are either myelinated or unmyelinated and are less than 6 μ in diameter.

These fibers possibly transmit "touch" when gently deformed or irritated and transmit pain when violently traumatized or irritated by metabolites of cell injury, ischemia, or inflammation. Within a population of finely myelinated fibers or unmyelinated fibers there are fibers that respond only to intense mechanical or thermal stimulae.

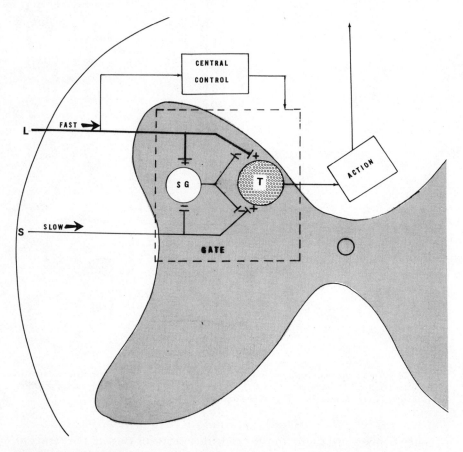

FIGURE 15.   Wahl-Melzak concept of gate theory of pain transmission.
SG — substantia gelatinosa   T — T cells

28

In recent years, as a result of the work of Wahl and Melzak, neurophysiologic investigation of pain has evolved a concept of a "gate" in the region of the dorsal horn in the spinal cord gray matter (Fig. 15). This concept attempts to clarify the role of specificity versus pattern.

The concept postulates that the faster-conducting large fibers, which carry mainly touch and proprioception, project to the substantia gelatinosa and then to the T cells. The substantia gelatinosa exerts an inhibitory influence upon the affected fibers. The faster fibers increase the substantia gelatinosa activity but the slower C fibers, allegedly related to pain transmission, exert inhibibitory influence. Since the slower impulses reach the gate later than the larger fiber impulses they find the gate closed. How the receptor endings and the slower fibers carry the sensation of pain is not clarified, but apparently various thresholds for basic stimuli exist.

Crue[3] refutes the old Aristotle concept of primary senses—

> vision
> hearing
> taste
> smell
> peripheral
>
> cutaneous   proprioception
>   touch
>   pressure
>   heat
>   cold
>   pain

—to a classification essentially based on physical energy source:

> electromagnetic
> chemical
> gravity
> mechanical displacement
> cutaneous sensibility
>   mechanical
>   thermal

and thus refutes noxiceptive receptors. Further neurophysiologic studies reveal that T cells are capable of independent firing. This is enhanced when they are sensitized by depolarization. The T cells are postsynaptic and can be potentiated from the periphery or from central sources. In acute pain the firing of the T cells is orderly and limited, whereas in chronic pain T cell impulses can be progressive, uncontrolled, and even epileptiform.

The central reactivity is markedly affected by the input of the peripheral impulses. The extent of this summation is not yet validated. It is apparent that pain is not influenced exclusively by the intensity of afferent fiber input but is influenced markedly by higher central processes (Fig. 16). Clinically, there is still need to therapeutically influence the afferent input of the peripheral fibers until researchers can clarify central control and its influence by therapeutic measures, whether they are psychologic, electric, thermal, or pharmaceutic.

Deep tissue sensibility is not as well documented and there is poor correlation between pain originating from a deep tissue to the superficial area supplied by that same spinal nerve.

Pain can be mediated via the posterior primary division as well as the anterior primary division. Thus, muscle spasm, which may result from a noxious stimulus, may be the primary site of irritation that is referred

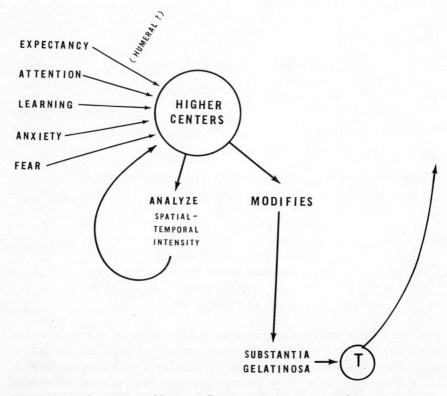

FIGURE 16. Higher center modification influencing gate transmission of pain.

down the same segmental anterior primary division. Stimulation of the ventral root produces a pain along the myotome supplied by that nerve.

Of the numerous soft tissues that can be the locus of pain, muscle tissue is a prime example. *Spasm* has been claimed for centuries to occur in painful states to prevent motion of an inflamed organ or joint and has, in turn, also been claimed to be the site or cause of pain. *Cramps* fall into the same category as spasm and have received very little study. Unusual or unaccustomed physical activities cause *dull ache* in the muscle; this persists briefly or disappears when that activity becomes habitual. None of these sensations can be attributed to the end products of ischemia.

Prolongation of muscular contractions causes irritation of the splinting muscle and its associated ligaments and tendons; this causes further spasm and initiates a painful cycle. The duration of a sustained contraction influences the severity of pain. In a sustained muscle contraction, such as gripping, the contraction is initiated by a limited group of muscle bundles. As tension is maintained, another group contracts and the initial group releases.

Ischemia has been designated as the cause of muscle pain in conditions such as angina, intermittent claudication, and even in conditions such as occipital tension headaches resulting from tension in the involved muscles.

Variation of arterial pressure influences the severity and rapidity of onset of pain. For example, an exercise done with the arm elevated, a position which increases arterial pressure, causes earlier and more severe pain than an exercise done with the arm held horizontally.

Studies performed with muscles contracting during proximal application of a tourniquet have presented many interesting conclusions. Without a tourniquet, repeated muscular contractions ultimately cause pain and depend upon the rate of contraction and the length of interval periods of relaxation. This implies that pain is due to accumulation of catabolites of the muscle metabolism of contraction. With a tourniquet applied proximally to the exercising muscles, significantly fewer contractions can be performed before onset of pain, and pain subsides rapidly upon release of the tourniquet.

The concept that pain can be attributed to oxygen depletion has been rejected because merely applying a tourniquet for a long period, without simultaneous exercising does not cause pain nor does application before exercise cause pain upon contraction of the ischemic muscle group performed with the tourniquet removed.

Releasing the tourniquet for a brief period does not completely remove metabolites since exercises performed after removal of the tourniquet, but without reapplication of the tourniquet, allow only limited subsequent painfree exercises. This implies retention of the irritating catabolites within the tissues. After pain-producing exercises with an applied tourniquet, release of the tourniquet ultimately relieves the pain, but the reapplication of the tourniquet without exercise reproduces the pain. Apparently this pain is due to retained metabolites.

Accumulation of irritating metabolites is still considered the causation, but the specific metabolite has not been identified. The once considered idea that lactic acid or pyruvic acid are the factor has since been refuted because ischemic exercises by patients with hereditary absence of muscle phosphorylase (McArdle's syndrome) develop severe pain, more severe than average, and lactic acid cannot be produced.

Catabolites are apparently produced by muscle fibers and find their way to the extracellular fluid to be removed by adequate capillary circulation. When capillary circulation is decreased by sustained muscle contraction, the concentration of catabolites adjacent to nociceptors increases and initiates impulses through the central nervous system via whichever pathway may be implicated.

These findings imply that muscle pain and inability to contract are not due to a depletion of necessary metabolic substances as from fatigue but are due to production of an excessive catabolic end product that must be removed by adequate blood flow.

These concepts give credence to the diagnosis of painful states resulting from tension, muscle spasm, cramps, and so forth and to the validity of physical therapeutic modalities that decrease muscle tension, improve faulty neuromuscular contraction, and improve local tissue circulation.

Sciatica, always considered a "nerve radiating pain" from compression at the intervertebral foramen, was initially attributed to muscle spasm.[4,5] When spinal root movement at the intervertebral foramen was observed on straight leg raising, the sign was attributed to increased tension within the nerve. With sciatica becoming synonymous with herniated lumbar disk,[6] the positive straight leg raising (SLR) test became positive indication of pressure upon the root at the intervertebral foramen.

This assumption has been questioned recently[7] because the straight leg raising test failed to remain positive after interrupting the arc between the posterior primary division, the cord, and the anterior primary division of a root segment by injecting trigger points[8] or after interrupting the articular branch of the posterior primary division[9,10] or by acupuncture into the area of the erector spinal muscles.

Sensitive areas on the soft tissues throughout the body have been described for years. Pressure upon these areas have caused local pain and pain referred into distal areas of the body. These referral zones have varied from dermatomal areas to areas other than dermatomal and have been termed sclerodermal.

These tender areas have been classified by numerous labels such as trigger point, myofascial pain syndrome, myalgia, myositis, fibrositis, fibromyositis, fascitis, myofascitis, muscular rheumatism, strain, and sprain.

The small hypersensitive region that constitutes a trigger site may be stimulated by pressure, needling, excessive heat or cold applied locally, or by motion that shields the tissues on which the sensitive zone is located (Fig. 17).

Trigger zones were originally described in 1936 with reproduction of pain referred to the shoulder and down the arm from pressure over the upper area of the scapula. Steindler and Luck[11] localized the site of referred pain as a ligament or a muscle.

Travell[12] has reported many studies on trigger points since 1942. All her patients showed equisitively tender spots that, when pressed, referred pain to a distal site.

In 1843 Froriep[13] labeled muscles that developed bands or cord within their structure as *rheumatism.* He believed that the palpable changes in the muscle was a connective tissue deposit. Most of these deposits were painfully sensitive to pressure. Grauhan,[14] shortly thereafter, described the hard nodules histologically as fibrous connective tissues surrounding sparse muscle fibers with infiltration of lymphocytes. Since then the

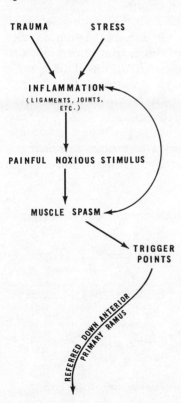

FIGURE 17.   Schematic concept of pain manifestation. Trauma to joints, ligaments, and other soft tissues ultimately creates trigger points that refer to distal sites.

literature has been inundated with similar discussions with labels such as fibrositis, interstitial myofibrositis, muscular rheumatism, nonarticular rheumatism, myofascial syndrome, myalgia, and myofascitis. These conditions have also been classified as acute, subacute, and chronic, with the acute being classified as those having severe pain and pressure sensitivity with reflex spasm, swelling, impaired mobility of the joints, and an increased temperature of the area. Most frequently involved have been the lumbar area, thigh, and sternocleidomastoid, trapezius, and levator scapulae muscles. The subacute phase is the condition in which pain and stiffness have decreased but are still present; the chronic phase is one in which the nodule is present without any symptomatology. Many of the nodules are reported as having persisted even during deep anesthesia of the patient, therefore eliminating the probability that myofascitis is a psychogenic condition exclusively.

Criteria for making a diagnosis of this disease entity was specified through the literature as (1) exquisite pain tenderness, (2) circumscribed painful hardening, and (3) pressure on the point that caused referred pain. In addition, however, to these criteria is the conclusion that muscle spasm which accompanies fibrositis is accompanied by muscular activities as is evidenced by specific electromyograph studies.

Interstitial myofibrositis has been postulated for this syndrome[15] with trauma considered the major causative factor (Fig. 18). Trauma varies from direct mechanical injury and eccentric muscular contraction to repeated muscular contraction caused by axone irritation. This latter is exemplified in axone irritation from osteophyte or herniated disk.

Local trauma causes edema as a resultant of microscopic herniation of muscle fascicles with outpouring of mast cells and platelets. The mast cells release mucopolysaccharides containing heparin and histamine, which cause vasodilatation; the platelets release serotoxin, which is a vasoconstrictor. The trauma causes a metachromatic mucoid substance to

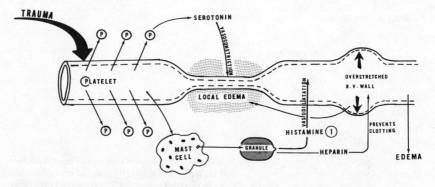

FIGURE 18.   Schematic concept of the effect of trauma, a vasochemical reaction.

infiltrate the interfascial and intercellular spaces, causing an edematous trigger area.

This concept of interstitial myofibrositis has merit and tends to clarify the benefits of many therapeutic approaches. Acetylsalicylic acid (ASA) inhibits blood platelets from releasing serotoxin; steroids inhibit the synthesis of mucopolysaccharides; heat causes vasodilatation to permit the removal of toxic catabolites; and massage and ultrasound mechanically irritate the mast cells to enhance degranulation. Nerve blocks intervene in the pain-spasm cycle and encourage vasodilatation.

Predisposing factors that create trigger sites are chronic muscular strain, repeated excessive muscular activity, direct trauma, chilling of fatigued muscles, various types of arthritis, nerve root injury, or psychogenic anxiety tension state.

Travell[16] reported benefit from spraying the trigger points with either ethyl chloride or vasocoolant spray. She advocates specific guidelines to this cooling spray technique: (1) a fine jet not a spray be used; (2) waft it across the involved skin in one direction; (3) cover no more than 4 inches per second of spraying; (4) cool the skin not frost it; and (5) accompany the spray with gentle passive stretching of the tissues containing the trigger site.

The entire literature of muscle pain syndrome has been thoroughly reviewed by Simon.[17]

Deep vessel injury or disease may be the noxious irritant that initiates the reflex arc. To mention one common example, myocardial infarction has been shown to cause chest muscle spasm. Interruption of this cycle by measures such as heat, electrical stimulation, needling, injections, or deep massage has resulted in relief of pain mediated from deep viscera or tissues.

Kraft and coworkers[18] claimed four criteria of myofascial pain syndromes: (1) the "jump sign" in which a patient jumps when the trigger area is palpated; (2) the rope muscle sign when muscle is carefully palpated (the sign is given great credence by Maigne[19] in musculoskeletal dysfunction); (3) excessive blanching followed by hyperemia upon shaking the area, as evidence of autonomic involvement; and (4) eradication of the syndrome by infiltration of trigger area with a local anesthetic or ethyl chloride spray.

Muscle spasm, whether the cause or merely an arc in the reflex, plays a major role in other painful syndromes. The motor unit normally is fired by the large alpha spinal motor neuron during voluntary effort, firing repetitively at rates up to 50/second. Each muscle fiber responds to a nerve impulse with a twitch (contraction). Each twitch is longer than the duration of the action potential, and so, with repeated action potentials known as tetanus, a smooth contraction results.

The number and rate of spinal motoneurons activated are controlled by excitatory and inhibiting synaptic mechanisms involving the fusimotor system. Muscle cramps with resultant stiffness may, therefore, occur from:

1. excessive descending facilitory impulses with greater alpha motoneuron activity
2. impaired inhibitory pathways that reflexly oppose alpha activity

These neurophysiologic mechanisms are not yet clinically accepted. Muscle spasm may also be related to failure of calcium to accumulate sarcoplasmic reticulum in the muscle or to abnormal reaction of actin and myosin, but these are not probable in the usual and frequent clinical manifestations of painful spasm.

Anxiety is often accompanied by muscular overactivity. This was postulated by Jacobson,[20] then verified by laboratory methods by Malmo and associates in 1948.[21] In 1952 Holmes and Wolf[22] found that patients with backache had generalized overactivity of the trunk muscles in situations which engendered conflict, insecurity, hostility, frustration, and guilt. These findings and others account for the often noted painful musculoskeletal symptoms attributed to tension, anxiety, etc., but occasionally they failed to differentiate which preceded the other.

Injuries causing severe sprain such as is incurred in athletic injury to the knee or ankle are essentially ligamentous and capsular. These injuries, however, are accompanied by swelling of the joint, a condition that implies perivascular edema and microscopic hemorrhage. This soft tissue exudate into fascial layers and into subcutaneous and periarticular areas is a deterrent to healing.

Currently the diagnosis of strain and sprain is made immediately by observation of a trainer, a coach, or a physician and elevation of the part with application of ice is standard treatment. Compression dressings to prevent further formation of hemorrhage or edema are also applied.

Since there may be arterial and, thus, capillary circulation to the injured part, occlusion of this arterial supply to the region would seem to be physiologically beneficial. Immediate application of a tourniquet by means of a sphygmomanometer, proximal to the injured joint, with pressure slightly exceeding arterial pressure would seem to be desirable. This has not been advocated as a recommended procedure.

Arterial occlusion for periods of 15 to 20 minutes would allow capillary closure by clotting, prevent edema of the injured part, and permit more careful examination of the part. Brief release of the sphygmomanometer followed again by reocclusion of the arterial supply would prevent tissue ischemia and resultant pain. A firm dressing then applied to the injured joint could be applied which would prevent swelling when the tourniquet is no longer applied. This recommendation is discussed further in the treatment of acute injuries in the chapters on the knee and ankle.

The study of pain originating in the neuromusculoskeletal system is not yet clarified. No one theory is universally accepted nor does any one concept confirm any specific painful disease entity. The musculoskeletal painful syndrome, albeit prevalent, is no more clear in all its ramifications than is pain and disability from any other organ system.

This chapter highlights what is known. At best much of the text is still postulation. It intends to give some basis for greater interest in soft tissue pain and disability and hopefully to direct interest to better understanding and management.

When a patient's complaint of musculoskeletal pain is attributed to faulty biomechanics, misuse or abuse, or psychologic causation without benefit of treatment, an organic illness not originally suspected nor clearly understood may be present. Polymyalgia rheumatica is such an example.

Our ignorance should never lead to complacency nor should any patient who complains of a disabling musculoskeletal pain be denied a searching mind nor a willingness to explore innovative treatment.

POLYMYALGIA RHEUMATICA

In middle-aged or elderly patients who complain of proximal muscle aching and stiffness, the condition of polymyalgia rheumatica must be considered. This is a potentially serious illness that responds dramatically to steroid therapy and yet can mimic and be confused with numerous other illnesses.

Polymyalgia rheumatica per se can be an innocuous illness yet may be related to giant cell arthritis, which can be serious. The latter must constantly be considered when the former is diagnosed and being treated.

Polymyalgia rheumatica without arteritis reveals no morphologic abnormalities and muscle biopsy is *not* revealing nor diagnostic. In the arteritis complication there is inflammation of the entire thickness of the blood vessel wall. The media smooth muscle undergoes necrosis, the intima proliferates, the internal elastic membrane fragments, and the adventitia thickens. This infiltration into the blood vessel by giant cells and macrophages occlude the lumen. All size arteries except the smaller arterioles can be affected.

Polymyalgia usually appears in women (2:1) of middle age (55 or older). A diagnosis in a person less than 50 should be considered.

Onset usually is insidious and consists of muscular pain and aching. Proximal muscles are involved, predominantly of the neck, shoulders, back, buttocks, and thighs. The discomfort is usually symmetric and is muscular rather than articular or tendinous.

Pain may be minimal at rest but is aggravated by any exertion. Muscle stiffness is prevalent especially upon arising in the morning or after periods of inactivity. Sleep is frequently disturbed.

Tenderness is infrequent and muscle weakness is rare. However, frequently there are associated symptoms such as fatigue, lethargy, weight loss, malaise, and depression. These associated symptoms gradually lead to disability and imply also a psychologic disease state.

The physical examination of the joints and muscles *usually is completely normal.* There is *no* muscle weakness or atrophy and little or no muscle tenderness. Joints have full range of motion. Frequently a diagnosis of functional illness is made because of the severe nonspecific complaints and a normal physical examination.

The most important diagnostic laboratory test is a markedly elevated erythrocyte sedimentation rate (ESR). The patient may have a mild anemia but muscle biopsy and muscle enzyme studies are normal. Response to steroid therapy is diagnostic as well as therapeutic.

Polymyalgia rheumatica is self-limited, usually lasting from two to six years. Treatment during the active phase consists of corticosteroids beginning with small doses of 10 mgm. daily or every other day and gradually increasing until symptoms subside and sedimentation decreases to normal. Dosage must be decreased gradually to maintain good clinical response and normal sedimentation rate. This controls the disease and minimizes the complications of long-term steroid therapy.

The complications of giant cell arteritis are always to be considered and treatment must be energetic to prevent the serious sequelae of the arteritis. On examination the palpable arteries (for instance the temporal artery) are tender and nodular and pulses are diminished or absent. Claudication, headaches, and visual impairment are examples of vascular insufficiency. In this complication initial large doses of steroids (40 to 60 mgm. daily) may be necessary and gradually decreased after seven to ten days.

Polymyalgia rheumatica must be differentiated from polymyositis, rheumatoid arthritis, lupus, scleroderma, or latent malignancy.

REFERENCES

1. Steinbach, R. A., et al.: Chronic low back pain: "low-back loser." Postgrad. Med. 53:135–138, 1973.
2. Bonica, J. J.: The Management of Pain. Lea & Febiger, Philadelphia, 1953.
3. Crue, B. L.: Pain. The Bull., Los Angeles Co. Med. Assoc., Oct. 4, 1973, pp. 10–15.
4. Forst, J. J.: Contribution a l'etude clinique de la sciatique. Paris These, No. 33, 1881.
5. Lasegue, Ch.: Considerations sur la sciatique. Arch. Gen. Med. 2 (Serie 6, Tome 4):558–580, 1864.

6. Minter, W. J., and Barr, J. S.: Rupture of the intervertebral disk with involvement of the spinal canal. N. Engl. J. Med. 211:210, 215, 1934.
7. King, J. S., and Lagger, R.: Sciatica viewed as a referred pain syndrome. Surg. Neurol. 5:46–50, 1976.
8. Travell, J., and Rinzler, S. H.: The myofascial genesis of pain. Postgrad. Med. 11:425–434, 1952.
9. Rees, W. S.: Multiple bilateral subcutaneous rhizolysis of segmental nerves in the treatment of the intervertebral disc syndrome. Ann. Gen. Pract. 26:126–127, 1971.
10. Shealy, C. N.: Facets in back and sciatic pain: A new approach to a major pain syndrome. Minn. Med. 57:199–203, 1974.
11. Steindler, A., and Luck, J. V.: Differential diagnoses of pain in the low back. J.A.M.A. 110:106–113, 1938.
12. Travell, J., et al.: Pain and disability of shoulder and arm; treatment by intramuscular infiltration with procaine hydrochloride. J.A.M.A. 120:417, 1942.
13. Froriep: Ein Beitrag zur Pathologic and therapie des rheumatesmus. Weimar, Germany, 1843.
14. Grauhan, M.: Überden anatomeschen Befience bel einem Fall von myositis Rheumatica. Doctoral dissertation. Cassel, Weber, and Weidemeyer, 1912.
15. Awad, E. A.: Interstitial myofibrositis: Hypotheses of the mechanism. Arch. Phys. Med. Rehabil. 54:449–453, 1973.
16. Travell, J.: Ethyl chloride spray for painful muscle spasm. Arch. Phys. Med. 33:291–298, 1952.
17. Simon, D. G.: Muscle pain syndromes. Special review. Am. J. Phys. Med. Williams & Wilkins Co., Baltimore. Vol. 54, No. 6, 1975 Part I; Vol. 55, No. 1, 1976, Part II.
18. Kraft, J. H., Johnson, E. W., and LeBan, M. N.: The fibrositis syndrome. Arch. Phys. Med. 49:155–162, 1968.
19. Maigne, R.: Medical Orthopedics. Charles C Thomas, Springfield IL, 1975.
20. Jacobson, E.: Electrical measurements of neuromuscular states during mental activities. I. Imagination of movement involving skeletal muscles. Am. J. Physiol. 91:567–608, 1930.
21. Malmo, R. B., and Shagass, C.: Psychosomat. Med. 11:9, 1949, and 11:25, 1949.
22. Holmes, T. H., and Wolff, H. G.: Life situations, emotions, and backache. Psychosomat. Med. 14:18, 1952.

BIBLIOGRAPHY

Altschule, M. D.: Emotion and skeletal muscle function. Med. Sci. 11:163–164, 1962.
Anrep, G. J., Saalfeld, E. V.: The blood flow through the skeletal muscle in relation to its contraction. J. Physiol. 85:375–399, 1935.
Barcroft, H., Millen, J. L. E.: The blood flow through muscle during contraction. J. Physiol. 107:518–526, 1948.
Berges, P. V.: Myofascial pain syndromes. Postgrad. Med. 53:161–168, 1973.
Bowsher, D.: Termination of the central pain pathways in man: The conscious appreciation of pain. Brain 80:606–621, 1957.
Clark, W. C., and Hunt, H. F: Pain. In Downey, J. A., and Darling, R. C. (eds.): Physiological Bases of Rehabilitative Medicine. W. B. Saunders Co., Philadelphia, 1971, pp. 373–401.
Cobb, C. R., et al.: Electrical activity in muscle pain. Am. J. Phys. Med. 54:80–87, 1975.
Crue, B. L., and Carregal, E. J. A.: Post synaptic repetitive neurone discharge in neuralgic pain. Presented at International Symposium on Pain, May 1975, Seattle. To be published by Raven Press.
Feinstein, B., et al.: Experiments of pain referred from deep somatic tissues. J. Bone Joint Surg. 36[Am]:981–997, 1954.
Fordyce, W. E.: Operant conditioning as a treatment method in management of selected chronic pain problems. Northwest. Med. 69:580, 1970.
Gordon, G.: Role Theory and Illness: A Sociological Perspective. College & University Press, New Haven CN, 1966.
Hirschfeld, A. H., and Behan, R. C.: The accident process. J.A.M.A. 186:193–199, 1963.
Holmes, T., and Rahe, R.: The social readjustment rating scale. J. Psychosom. Res. 11:213–218, 1967.
Layzer, R. B., and Rowland, L. D.: Cramps. N. Engl. J. Med. 285:31–40, 1974.

Livingston, W. K.: Pain Mechanisms. MacMillan Co., New York, 1943, pp. 128, 139.

Lundervold, A.: Electromyographic investigation during sedentary work. Br. J. Phys. Med. 14:32–36, 1951.

Mead, G. H.: Mind Self and Society. University of Chicago Press, Chicago, 1952.

Mehler, W. R.: Some observations on secondary ascending afferent systems in the central nervous system in pain. Henry Ford Hospital International Symposia, Little Brown & Co., Boston, 1964, pp. 11–32.

Melzak, R., and Wall, P. D.: Pain mechanisms—a new theory. Science 150:971, 1965.

Neufeld, I.: Mechanical factors in the pathogenesis, prophylaxis and management of "fibrositis" (fibropathic syndromes). Arch. Phys. Med. Rehabil. 759–765, 1955.

Park, S. R., and Rodbard, S.: Effects of load and duration of tension on pain induced by muscular contraction. Am. J. Physiol. 203:735–738, 1962.

Parsons, T.: The Social System. Free Press, Glencol IL, 1951.

Perl, E. R.: Mode of action of nociceptors. In Hirsch, C., and Zotterman, Y. (eds.): Cervical Pain. Pergamon Press, Oxford, 1972.

Sainsbury, P., and Gibson, J. G.: Symptoms of anxiety and tension and the accompanying physiological changes in the muscular system. J. Neurol. Neurosurg. Psychiatry 17:216–224, 1954.

Simons, D. G.: Muscle pain syndromes, Part I. Am. J. Phys. Med. 54:289–311, 1975.

Stillwell, D. L.: The innervation of tendons and aponeurosis. Am. J. Anat. 100:289–317, 1957.

Travell, J.: Pain mechanisms in connective tissue. In Regan (ed.): Conference on Connective Tissue, Josiah Macy Jr. Foundation, New York, 1951, pp. 86–125.

Travell, J., and Rinzler, S. H.: The myofascial genesis of pain. Postgrad. Med. 11:425–434, 1952.

Walter, A.: The psychogenic regional pain syndrome and its diagnosis. In *Pain*. Little Brown & Co., Boston, 1966, Ch. 34.

CHAPTER 3

Low Back Pain

Of the numerous musculoskeletal disabling conditions, the complaint of low back pain is undoubtedly predominant. Estimates of hours lost to industry, hours of disability, and money paid for medical care and disability compensation is astronomic. There are as many concepts of the mechanisms and causes of low back pain as there are advocates of numerous forms of treatment. Currently there is no one accepted mechanism that has full acceptance and credence, nor can any one form of treatment be considered predominant and totally applicable to all complainers of low back pain.

A fundamental knowledge of functional anatomy of the anatomic part involved is mandatory before all the aspects that may impair the function of this anatomic site with resultant pain can be fully understood.

FUNCTIONAL UNIT

The vertebral column, which in its lumbar area is the site of pain, can best be evaluated by understanding of the functional units that comprise the vertebral column (Fig. 19). The functional unit is composed of two segments: the anterior portion, which is the weight-bearing structure, and the posterior portion, which functions in directional guidance (Fig. 20).

Anterior Portion

The weight-bearing portion of each functional unit is the anterior portion, which is comprised of two vertebral bodies separated by a hydrodynamic shock absorber, the intervertebral disk.

The intervertebral disk is a fibrocartilaginous body securely connecting two adjacent vertebral bodies. The individual disk comprises three parts:

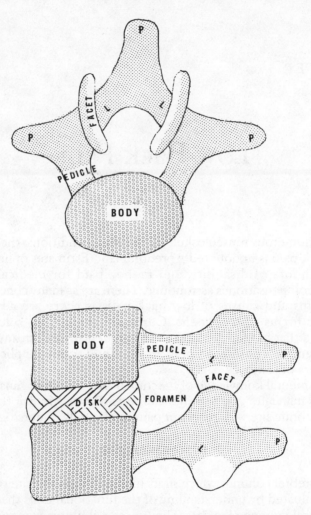

FIGURE 19. Functional vertebral unit. *Above,* View of the vertebral body, the posterior articulations (facets), the pedicles, the processes *(P),* and the lamina *(L). Below,* Lateral view demonstrating the intervertebral disk and its relationship to the components of the unit.

1. Hyaline cartilage plates at each end of the adjacent vertebral bodies.
2. The annulus fibrosus (Fig. 21) composed of concentric lamellae running obliquely between adjacent vertebral bodies and firmly attached to the circumference of the end plates. The annular fibers are enmeshed within a mucopolysaccharide matrix. The annular ring is thicker anteriorly and laterally than the posterior portion.
3. A centrally located pulpy nucleus (nucleus pulposus) (Fig. 21).

42

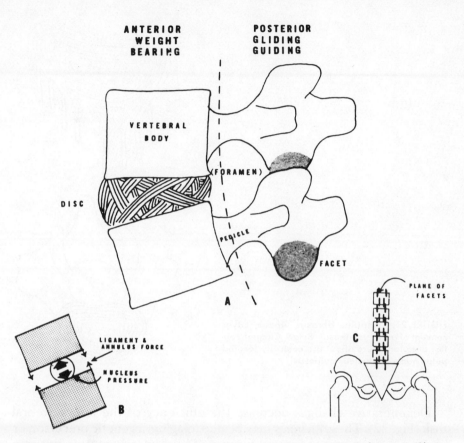

FIGURE 20. The functional unit of the spine in cross section: A, Lateral view; B, Shows pressure within disk, which forces vertebrae apart, and the balancing force of the long ligaments; C, Gliding motion of the plane of the facets.

The matrix of the disk contains 88 percent water. It varies from 88 percent at birth to 70 percent in senescence. The ground substance of the matrix is collagen and protein polysaccharide with a high proportion of chondroitin–14–sulfate and small amounts of hyaluronic acid. From birth through childhood the nucleus becomes more oval and increases in size. Around 21 years of age, the ring epiphysis of the vertebral bodies fuse and the annular fibers become firmly attached to the bone.

Aging plus multiple microtrauma causes the cartilage plates to become thinner and the posterior aspect of the annulus to become fragmented. There is some invasion of granulation tissue from the vertebral body along with gradual loss of water and the nucleus loses its ability to bind water; thus intradiskal pressure decreases. Further aging or trauma causes a decrease of protein polysaccharide concentration and the disk becomes more fibrous, inelastic, and inert hydrodynamically.

43

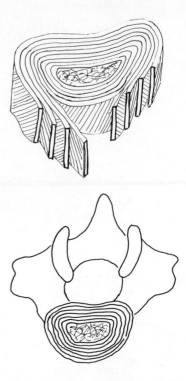

FIGURE 21. Annulus fibrosus. *Above,* Layer concept of annulus fibrosus; *Below,* Circumferential annular fibers about the centrally located pulpy nucleus (nucleus pulposus).

Degenerative changes decrease the efficiency of the intervertebral disk (Fig. 22). These changes may be due to aging, a genetic precursor, or chemical or microtraumatic factors. The hydrodynamics become impaired. The clinicopathologic significance of these changes will be discussed later in greater detail.

The disk functions as a hydraulic shock absorber. The pressure within the nucleus pushes the vertebrae apart and the annular fibers pull them together (Fig. 23). As weight is added to the unit, the vertebral bodies approximate by deforming the nucleus. Upon release of the compressing force, the nucleus regains its resting form. Flexion, extension, and some rotation is permitted by this nucleus deformation. Compression tests have confirmed that forces will fracture the vertebrae before damaging the disk.

The posterior longitudinal ligament reinforces the posterior aspect of the intervertebral disk. In the lumbar region the ligament tapers to become partial and, thus, affords inadequate protection to the contents of the spinal canal and the intervertebral foramina (Fig. 24).

Flexion, extension, and rotation is permitted because of the obliquity and the elasticity of the annular fibers (Fig. 25). Rotation of the vertebral column is restricted by the limit of extensibility of the annular fibers.

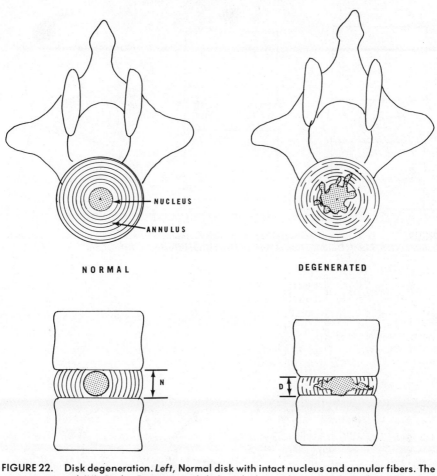

NORMAL   DEGENERATED

FIGURE 22. Disk degeneration. *Left,* Normal disk with intact nucleus and annular fibers. The space is normal *(N). Right,* Degenerated disk with the nucleus outside its boundary and fragmented annular fibers and narrowed space *(D).*

Excessive rotation exceeding this limit may be a contributing force to disk damage, herniation, and deterioration (Fig. 26).

Nutrition of the intervertebral disk remains an enigma. In the neonatal period many small blood vessels penetrate the vertebral end plates and supply the disk. These blood vessels are obliterated in adolescence and the disk becomes avascular. Nutrition of the disk by osmosis has been disproved and the current accepted method of nutrition is by imbibition. This forms a colloid imbibition pump.

As the disk imbibes, its size increases and causes the annular fibers to become taut. Equilibrium is reached by the hydraulic pressure exerted by the vertebral end plates above and below and by the encircling annular fibers.

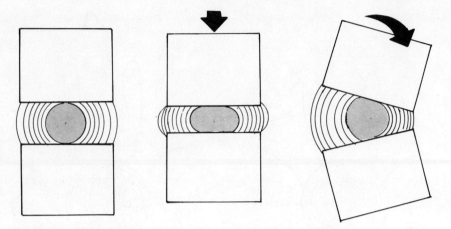

FIGURE 23. *Left*, Normal nonweightbearing disk; *Center*, Deformation of nucleus reacting to compression; *Right*, Deformation of nucleus permitting flexion or extension.

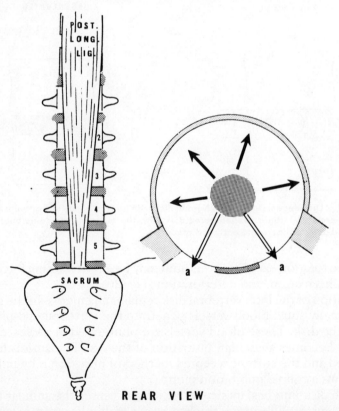

REAR VIEW

FIGURE 24. Inadequacy of the posterior longitudinal ligament in the lower lumbar segment, therefore decreasing the protective effect in the L4, L5, and S1 region. *Right*, Disk herniation may bulge into the spinal canal, *a*.

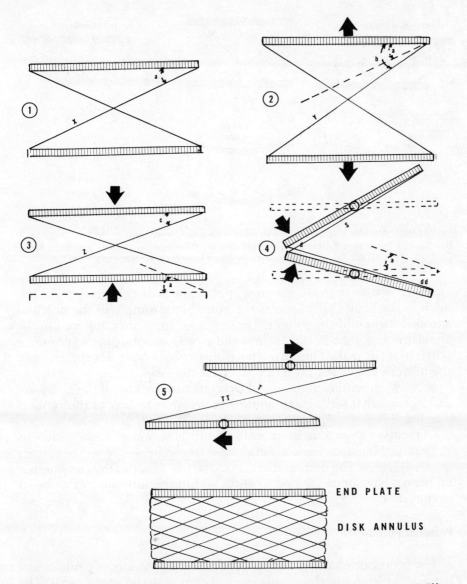

FIGURE 25. *1*, Annular fibers at rest; *2*, Effect of elongation; *3*, Effect of compression; *4*, Effect of flexion or extension; *5*, Effect of translatory torque.

Diffusion of solutes occurs via the central portion of the end plates and through the annulus (Fig. 27). Increased intradiskal pressure probably also forces fluid through minute foramina in the end plates. When pressure is released or decreased, fluid returns into the disk by imbibition.

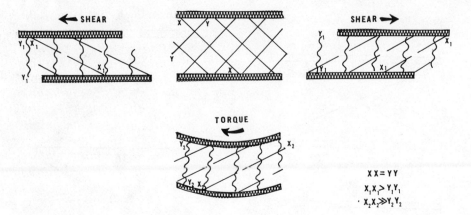

FIGURE 26. Annular fiber reaction to shear and torque. *Above,* Shear to the left elongates X fibers, thus permitting Y fibers to undulate and shorten; *Middle,* Normal positions; *Right,* Contrary shear effect; *Below,* Effect on annular fibers in rotatory torque.

There are numerous marrow spaces immediately under the endochondral plate that connects the bone blood supply to the disk and allows diffusion of the solutes. These spaces are more numerous in the annular region than in the nucleus and more numerous in the posterior portion of the disk. This factor possibly explains early degeneration of the nucleus and of the posterior aspect of the disk.

Solutes containing glucose and oxygen enter by way of the end plate and sulfates that form glucosaminoglycans enter by way of the annulus (see Fig. 27).

Currently the disk is not considered to have innervation within its substance. Numerous investigators have traced nerves of various stages of myelinization and demyelinization to the periphery of the annulus, but no nerves have been verified to enter within the substance of an intact normal disk.

Posterior Portion

The posterior elements of the functional unit consists of pedicles and laminae which form the circular arch comprising the spinal canal. The posterior spinal arch contains the posterior articular joints of the functional units (Fig. 28).

These facets are composed of articular cartilages on their opposing surfaces and have a capsule, synovium, and synovial fluid. By their shapes and verticality they articulate on each other to permit specific directions of motion and to deny or modify opposing directions of motion. Because of their sagittal vertical plane in the lumbar spine, the facets permit flexion and extension but deny or restrict lateral flexion and rotation (Fig.

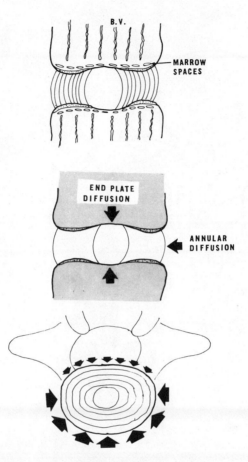

FIGURE 27. Disk nutrition through diffusion. Diffusion of solutes occurs through the central portion of the end plates and through the annulus. Marrow spaces exist between circulation and hyaline cartilage and are more numerous in the annulus than in the nucleus. Glucose and oxygen enter via the end plates. Sulfate to form glucosaminoglycans enters through the annulus. There is less diffusion into the posterior annulus. (B.V.— blood vessels)

29). Compared to the anterior joint of the intervertebral disk space, they are not weight bearing. It has been considered that they can support approximately 10 to 12 percent of body weight in the extended spine decreasing to no weight bearing with any significant degree of forward vertebral flexion. In lumbar hyperextension, lateral flexion and rotation can be completely prevented in this posture. The vertebral facets are in complete opposition in a "locked" position. With any degree of forward flexion, separation of the facets occur, thus permitting some degree of lateral flexion and some degree of rotation.

The posterior wall of the spinal canal, the anterior wall of the pedicles, and the lamina are covered by the yellow ligament (ligamentum flavum).

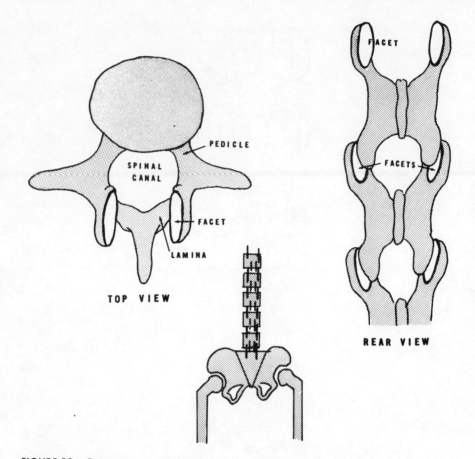

FIGURE 28. Posterior view of the lumbar spine depicting the vertical alignment of the posterior articulation (facets) and depicting the small musculature of the lumbar column.

The yellow ligament is an elastic longitudinal ligament extending the length of the vertebral column comprised exclusively of yellow elastic fibers. Its function has been considered to be prevention of the redundant capsule from becoming impinged between adjacent joint surfaces during extension of the spine and limitation of capsular bulging into the spinal canal during other movements of the vertebral column.

Completing the posterior arc of the vertebral body are the transverse processes and the posterior superior spine upon which attach supporting intervertebral ligaments and the intervertebral muscles that activate vertebral column motion.

The intervertebral foramina are formed by two superior and inferior adjacent pedicles. Anteriorly are the vertebrae bodies, the intervertebral disks, and the posterior longitudinal ligament. Posteriorly are situated

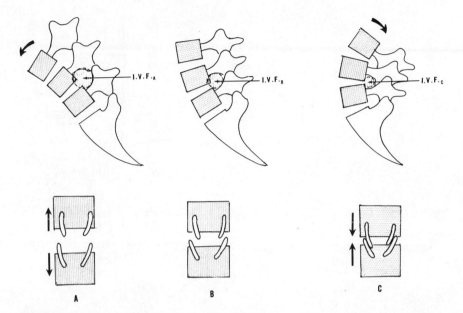

FIGURE 29. Facets in the lumbar spine. *A,* Separation of the facets in forward flexion; *B,* Opposition of the facets in the physiologic lordotic posture; *C,* Approximation and opposition of the facets on extension and hyperextension.

the facets, their capsules, and the yellow ligament. Through these foramina emerge the nerve roots, their dural sleeve, and the recurrent nerve of Luschka. The nerve roots descend at the cauda equina.

Nerve System in Lower Back

As the nerves descend the spinal canal they cross the disk immediately above the foramen. They enter the foramen beneath the pedicle. After the nerves leave the foramen they incline downward, outward, and forward and enter the origin of the psoas muscle by the time they reach the level of the lower disk (Fig. 30).

Whereas in the cervical spine the roots emerge horizontally in the lumbar region, here they oblique to emerge almost vertically. Because of their vertical entry into the foramen they are located in a superior position at the point of entry (Fig. 31).

Ventroflexion of the neck (cervical spine) tenses the lumbar and sacral nerves. This fact explains the accentuation of a positive straight leg raising (Lasegue) test by simultaneously flexing the neck.

The nerve roots within the foramen invaginate the dura and arachnoid, thus forming a funnel-shaped depression. Each nerve root carries a sheath of dura and arachnoid (Fig. 32). The arachnoid continues along

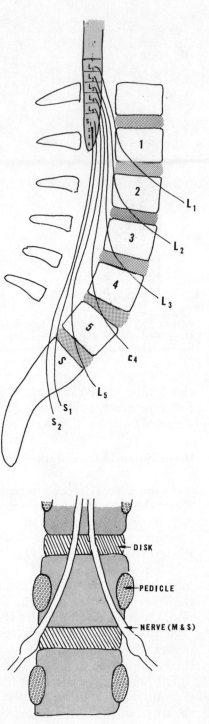

FIGURE 30. The nerve roots emerge from the conus of the cord located at L₁ level and oblique downward to emerge below the correspondent vertebral body. The length of L₅ and the sacral roots are evident and the obliquity of each nerve root is also depicted.

FIGURE 31. Nerve root relationship to pedicles. The nerve descends the spinal canal and obliques out, passing the pedicle and crossing the inferior disk at its superior lateral aspect. There is little exposure of the nerve to the disk at this level. Most of the exposure is in the immediate cephalad disk.

52

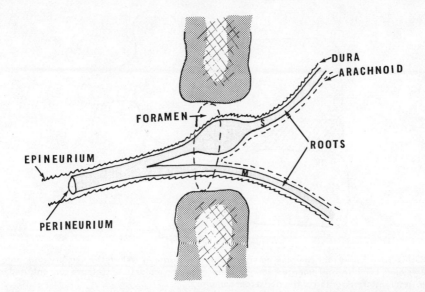

FIGURE 32. Dural and arachnoid sheaths of the nerve root complex. The arachnoid follows the sensory and motor nerve roots to the beginning of the intervertebral foramen and follows the sensory root to the beginning of the ganglion. The dura follows the nerve roots until they become the combined sensory and motor nerve outside the foramen and continues as the perineurium and epineurium. Neither the dura nor the arachnoid attach to the intervertebral foramen.

the nerve root as far as the ganglion, but does not envelop it completely. The dura continues along the roots until they merge; then the dura continues along the combined nerve to form an outer fibrous sheath, the perineurium. The epidural tissue (epineurium) continues along with the perineurium to reinforce the nerve sheath.

The posterior nerve (sensory) root is twice the thickness of the anterior (motor) root in the lumbar region. They remain separate as far as the ganglion and then merge into a single funiculus. The motor fibers are concentrated anteroinferiorly in the funiculus (Fig. 33).

Each dorsal root is composed of myelinated and unmyelinated nerve fibers. The large myelinated nerves conduct rapidly and transmit impulses from special peripheral receptors. The smaller myelinated and the very small unmyelinated fibers conduct slowly and transmit impulses from less specialized peripheral receptors.

The ventral roots, originally believed to contain only myelinated fibers, are now known to contain many unmyelinated fibers.

The nerve complex (roots and their investing sleeves) are not attached to the wall of the foramen and, thus, permit movement and tension within the foramen, but there is a limit to their elasticity. The nerve root fibers are arranged in parallel bundles with a thin endoneural sheath but *no* epineural or perineural tissue. The tensile strength of the peripheral

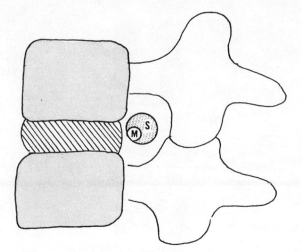

FIGURE 33. Nerve root complex emerging through foramen. When the sensory and motor nerve roots combine and emerge through the foramen, the motor fibers (M) are concentrated in the anteroinferior portion of the nerve root complex.

nerve is greater than that of the roots, which explains the damage to the roots in traction injuries.

Within the foramen the nerve and its investing sheath occupies 35 to 50 percent of the space, thus having ample space. The remainder of the foramen is filled with loose areolar connective tissue and adipose tissue, through which run the spinal artery, numerous veins, lymphatics, and the recurrent meningeal nerve.

Traction upon the nerve pulls the entire complex outward. Since the dura is cone-shaped, lateral motion pulls the foramen and resists further motion outward through the foramen (Fig. 34). In the cervical spine the epidural sheath is securely attached to the vertebral column.

If the disk were narrowed, the intervertebral foramen is narrowed as the superior articular process of the inferior vertebra moves upward, thus encroaching upon the contents of the foramen.

Although many of the related tissues can encroach upon the foramen, there is adequate space there before the nerves are encroached upon, implying that friction on the nerve as it passes may be a large factor on causation of symptoms.

There are surprisingly few studies on clinical or experimental pathologic studies of nerve root compression. Most studies in this area have been done on peripheral nerves.

PHYSIOLOGIC CURVES

The entire vertebral column comprises the supraincumbent functional units. The vertebral column has seven cervical vertebrae, twelve thoracic

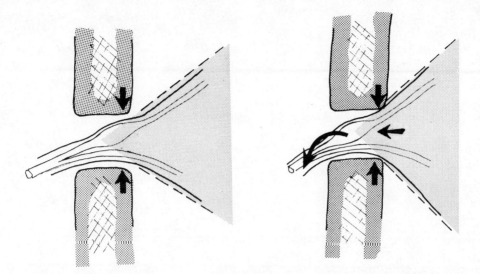

FIGURE 34. Dural protection to nerve root traction injury. *Left,* The dura forms a funnel which keeps the nerve root complex free within the foramen; *Right,* Effective traction draws the nerve root and its dura into the foramen, where it becomes blocked mechanically at the site of the arrows, making further traction impossible.

vertebrae, and usually five lumbar vertebrae. Curves are physiologically present in this vertebral column. A forward convexity in the lumbar and cervical spinal column forms the lordosis and an opposite curving with posterior convexity, known as kyphosis, is noted in the thoracic and sacral spinal components.

These physiologic curves are not present at birth (Fig. 35). The newborn has a spinal column with a total kyphotic curve, which probably is due to the absence of intervertebral disks.

As the child develops, elevation of the head initiates cervical lordosis and full extension of the vertebral column in the erect position gradually creates lumbar lordosis. This latter curvature of the lumbar spine is assumed to be partially caused by the failure to fully elongate the iliopsoas musculature.

In the erect posture the entire vertebral column is supported on an oblique sacral base that oscillates between the two hip joints. The angle of the superior surface of the sacrum is termed the lumbosacral angle (Fig. 36).

Upon this sacral base balances the entire vertebral column with all its physiologic curves. All curves are dependent upon the lumbosacral angle to retain its balance to the center of gravity. Increase of one angle demands an increase of the other (Fig. 37).

There are varying degrees of lumbosacral angle, dependent upon

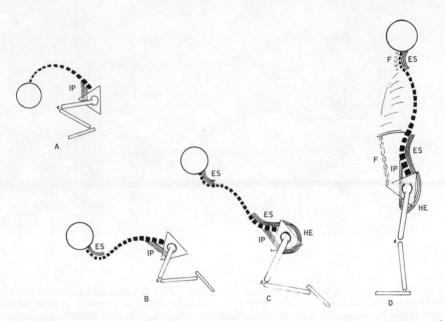

FIGURE 35. Evolution of erect posture. *A*, Fetal position with total kyphosis; *B*, Extension of cervical spine with elevation of head; *C*, Extension of lower extremities beginning stretch of iliopsoas muscle, causing lumbar lordosis; *D*, Full erect spine with cervical lumbar lordosis and dorsal and sacral kyphosis.

F —flexors
ES —extensors
IP —iliopsoas
HE—hip extensors

cultural, genetic, and racial differences. Changes in this angle are also caused by faulty habits, ligamentous laxity, and muscular tone.

ERECT BALANCE AND STANCE

Erect balance requires biomechanic balance. Since the vertebral column is flexible, it would appear that muscular effort is necessary to maintain column relationship to the center of gravity (Fig. 38). If this were true, muscular fatigue and collapse of the erect column would ensue. Nature has prevented this total dependence on muscular balance by permitting the column to be supported on ligamentous structure. The lumbar curve extends into greater lordosis, placing its support upon the anterior longitudinal ligament and posteriorly upon the facet joints (Fig. 39).

The hip joints can be immobilized in the extended position by virtue of the ligamentous bands within the anterior capsule of the hip joint. With the body weight "leaning" against this iliopectineal band (the Y ligament of Bigelow) the hip can remain extended with no muscular

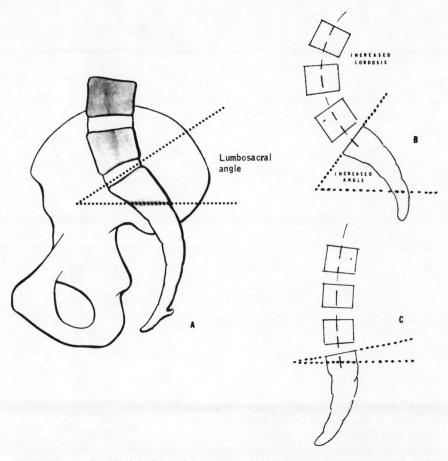

Lumbosacral angle

INCREASED LORDOSIS

INCREASED ANGLE

B

A

C

FIGURE 36. Lumbosacral angle. A, The lumbosacral angle is formed by a line paralleling the top of the sacrum and a line drawn horizontally. The spine is balanced upon the sacrum. As the angle increases *(B)* the lordosis is greater. With a lesser angle *(C)* the lumbar lordosis is less.

activity. The knee joint is prevented from hyperextending by virtue of the posterior popliteal capsular tissues. The knee can thus be maintained in the extended position by ligamentous support requiring no muscular effort.

Only the ankle cannot be immobilized and supported by ligamentous tissue. The ankle, however, can be maintained in the erect weight bearing position with alternating isometric contraction of the anterior dorsiflexors and the posterior gastrocnemius-soleus muscle groups (Fig. 40).

Since forward flexion of the lumbar spine is essentially only a reversal of the lumbar lordosis into a slight lumbar kyphosis, full body flexion requires simultaneous rotation about the hip joints of the pelvis as the

57

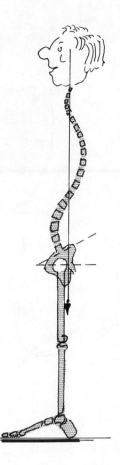

FIGURE 37. Erect balanced posture. The four
physiologic vertebrae curves balanced upon the sacral
base transect the plumb line of center of gravity.

lumbar spine flexes (Fig. 41). The synchronous lumbar spine flexion with
pelvic rotation is aptly termed lumbar-pelvic rhythm. For every degree of
lumbar spine flexion there is a proportional rotation of the pelvis.

Because the erect stance is essentially a posture dependent upon
ligamentous support, muscular support must be elicited the instant the
body flexes forward anterior to the center of gravity. As the body flexes
forward, the erector spinae muscles elongate to decelerate forward
flexion (Fig. 42). Insofar as muscular elongation which performs
deceleration is a learned reflex, its smoothness and efficiency requires
neuromuscular coordination. Poor training or distraction during this
motion will interrupt the smooth efficiency of forward flexion. Poor body
condition may also, by increasing the fibrous elements of the muscular
system, limit the degree of extensibility.

Once the body reaches full forward flexion it is again restricted by
ligamentous connective tissue.

To return to the full erect posture the extensor muscles must now

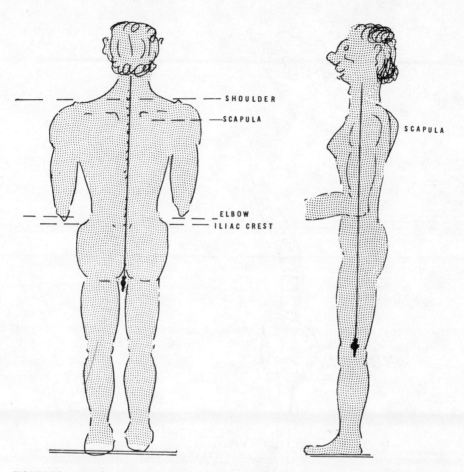

SHOULDER

SCAPULA

SCAPULA

ELBOW
ILIAC CREST

FIGURE 38.   Standing posture: center of gravity. *Left,* An anteroposterior view of the center of gravity shows the plumb line descending from the occiput through the sacrum. *Right,* A lateral view shows the plumb line passing through the cervical vertebrae, through the shoulder of the dangling arm, posterior to the hip joint, anterior to the center of the knee joint, and slightly anterior to the lateral malleolus of the ankle.

shorten and function as acceleration or body elevation until the full erect posture of ligamentous support is achieved.

The return to erect posture must also accomplish the reverse of the flexion lumbar-pelvic rhythm. The lumbar lordosis must be gradually resumed as the pelvis is *derotated.* Here again, precise coordinated neuromuscular effort must exist and requires training, practice, and no interference by distraction.

Insofar as the posterior articulations, the facets, guide the direction of lumbar flexion and extension and deny lateral flexion and rotation, this direction of reextension must be carefully observed.

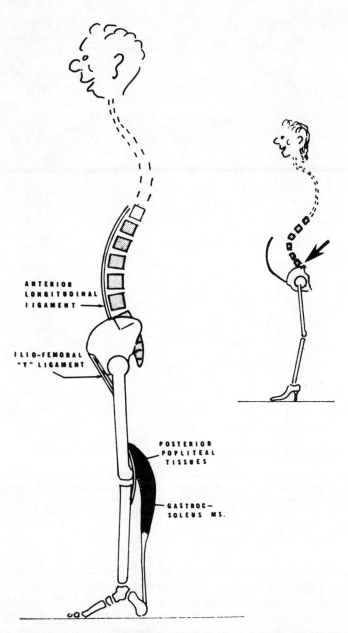

ANTERIOR
LONGITUDINAL
LIGAMENT

ILIO-FEMORAL
"Y" LIGAMENT

POSTERIOR
POPLITEAL
TISSUES

GASTROC-
SOLEUS MS.

FIGURE 39. Ligamentous erect balance. Erect posture can be maintained by ligamentous support. The anterior longitudinal ligament supports the lordotic-lumbar curve. The hip joints are supported by the anterior iliofemoral ligaments and the knee is locked by the posterior popliteal ligaments and capsule. Only the ankle cannot be kept inactive by ligamentous support.

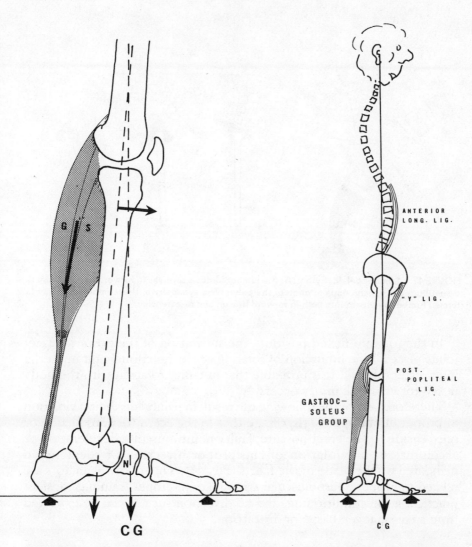

FIGURE 40. Ligamentous support of the erect non-moving spine with leaning of the lumbar spine on the anterior longitudinal ligament, the anterior hip joint on the iliopectineal ligament, and the knee on posterior popliteal ligaments. The gastrocnemius-soleus group depicted to the left contract isometrically, thus preventing the lower leg from leaning anteriorly. The erect stance is possible with no muscular action except intermittent sporadic isometric contraction of the gastrocnemius-soleus.

In the flexed position, i.e., the body bent forward and the lumbar lordosis reversed, the facets are separated, thus permitting some degree of lateral flexion and rotation. As lordosis is achieved, the facets approximate and, ultimately, as hyperextension is achieved, are in direct contact and prevent *any* lateral or rotation movement.

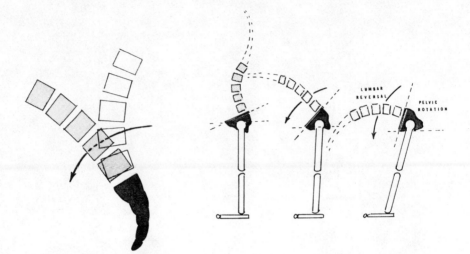

FIGURE 41.　Lumbar pelvic rhythm. As the lumbar lordosis reverses in forward flexion there is a simultaneous synchronous rotation of the pelvis. The symmetry of this combined motion is termed lumbar pelvic rhythm both in forward flexion and re-extension.

In the forward flexed position, the separation of the posterior facet joints place all the limitation of rotation of the functional unit upon the intervertebral disk. It is probable that in faulty rotation with the body flexed forward disk injury occurs.

Therefore, any of the following can result in painful disability: violation of proper lumbar-pelvic rhythm, either in the act of forward flexion or reextension to the erect posture; faulty neuromuscular deceleration or acceleration; or violation of the proper direction of reextension mandated by the plane of the facets (Fig. 43). The following factors are related to proper neuromuscular coordination: proper training, constant practice, good conditioning, and no diversion such as can be expected from anxiety, anger, haste, or distraction.

SITE OF PAIN

Since faulty action of the functional vertebral unit can result in pain, it is necessary to elicit the tissues within the unit which, when irritated, can initiate the sensation of pain. Within the functional unit, the following should be noted.

1. The anterolongitudinal ligament is sparsely innervated, but chemical, electrical, or mechanical irritation of this ligament does *not* evoke any significant pain sensation.

2. Vertebral bodies are a site of pain as is evidenced in fracture,

62

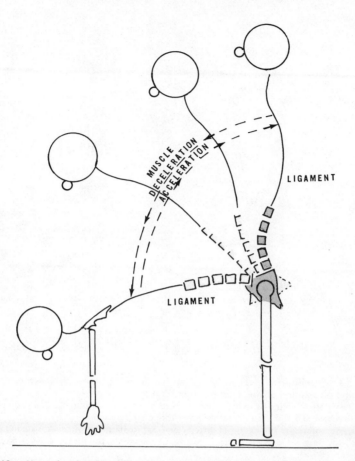

FIGURE 42. Muscular deceleration and acceleration of the forward flexing spine from the erect ligamentous support to the fully flexed ligamentous restriction. Muscle eccentric and concentric contraction permits forward flexion and re-extension.

metabolic disorder, or metastatic disease. The pain from the bodies themselves is dull, vague, usually nonradiating, and not significantly related to motion or position. Nocturnal pain in the reclining position is a symptom that alerts the examiner to suspect disease of malignancy or metabolic origin and initiates appropriate studies.

3. The young undamaged intervertebral disk with no evidence of degenerative changes is avascular, aneural, and, therefore, insensitive. No pain is elicited when performing a diskogram, which requires penetration of the pulpy nucleus with a needle. Injection of a small quantity of material without undue pressure causes no discomfort other than a mild low backache. Injection of an anesthetic agent such as Novocaine does not alter a residual low backache; therefore, intradiskal innervation is

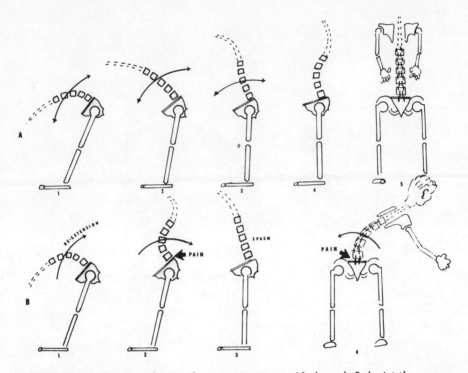

FIGURE 43. Mechanism of acute facet impingement. *A1* through *5* depict the proper physiologic resumption of the erect position from total flexion with reverse lumbar-pelvic rhythm. Re-extension must be in the anterior posterior plane as shown in *A5*. *B* shows improper premature return of lordotic curve before adequate pelvic derotation *(B2)*. This cantilevers the lumbar spine anterior to the center of gravity and approximates the facets, causing pain *(B2)*. With the body flexed and rotated there is further asymmetry of the facets facilitating unilateral impingement *(B4)*.

verified and altered dynamics must be assumed. Injection of material into an abnormal disk causes severe low back pain with radiation of leg pain.

4. The posterolongitudinal ligament has copious innervation of unmyelinated nerves with sympathetic fibers (Fig. 44). The nerve supply has been firmly established to be the recurrent meningeal nerve of Luschka. Pain can occur from irritation of these innervated tissues.

5. The nerve of Luschka also innervates the dural sheath of the nerve root as it emerges through the intervertebral foramen (Fig. 45). The nerve root and its dura are another source of pain within the functional unit. The intrinsic nerve supply from the meningeal ramus has several branches that run through the foramen: one towards the posterior longitudinal ligament, one to the dura mater, and one within the epidural tissue (Fig. 46). There are numerous fibers that run longitudinally on the dorsal surface or within the dorsal dura but none are found on the ventral surface or intradurally. The lack of nerve supply to the ventral aspect of

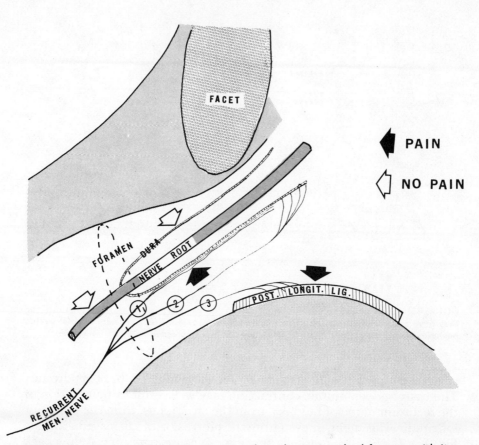

FIGURE 44. Dura accompanying nerve root through intervertebral foramen with its innervation by the recurrent meningeal nerve. *1* and *2* proceed to the anterior dural sheath, illustrating the sensitivity of that portion of the dura. The posterior dural sheath with no innervation is insensitive. *3,* Innervation which is capable of pain supplies the postero-longitudinal ligament.

the dura possibly explains Falconer and associates[1] in their claim that the dura was insensitive.

6. The yellow ligament is exclusively yellow elastic connective tissue and is devoid of any innervation; therefore it is insensitive.

7. The interspinous ligaments are also innervated and can, when inflamed, cause local as well as referred pain. Kellgren[2] injected irritating substance alongside of the spinous process of the first sacral vertebra and caused pain of a sciatica nature radiating down the leg. His injections apparently were into the ligament, partially into the muscle and may have been into the periosteum.

8. Stimulation of deep muscles around a spinal joint has been confirmed to cause referred pain. The multifid muscles contract in an

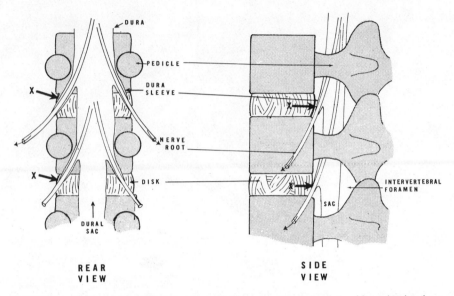

FIGURE 45. Dural sleeves of the nerve roots. As the nerve roots move lateral from the dural sac they carry a sleeve of dura through the intervertebral foramen. They pass near the disk in this posterior lateral region, and it is here that they can be compressed by a disk herniation.

attempt to splint the lumbosacral spine in painful low back syndromes. The basis of deep muscle contraction may well result from irritation of the recurrent nerve of Luschka, from the facet joints, from the posterior longitudinal ligament, or even from the protrusion of a damaged disk (see Fig. 46). If excessive muscular tension can cause symptoms of low back and referred leg pain it can, in part, explain aggravation of symptoms from anxiety, anger, tension, and even chilling as well as faulty use of the back.

9. The posterior articulations (facets) are synovial joints and are a site of pain from injury, inflammation, or misuse. The facets are innervated by the articular branch of the posterior primary division of the nerve root (Fig. 47).

In summarization, the tissues capable of causing pain are the posterior longitudinal ligament, the nerve root and its dura, the posterior articulations (facets), the ligaments, and the musculature of the vertebral column.

The examiner should appreciate and determine which of these tissues is the site of pain and the mechanism by which these tissues are being irritated, thus initiating the pain cycle. Assuming that pain mediates through peripheral nerves and ascends the cord, then the mechanism of chemical, electrical, thermal, or mechanical nociceptive stimuli, which is

66

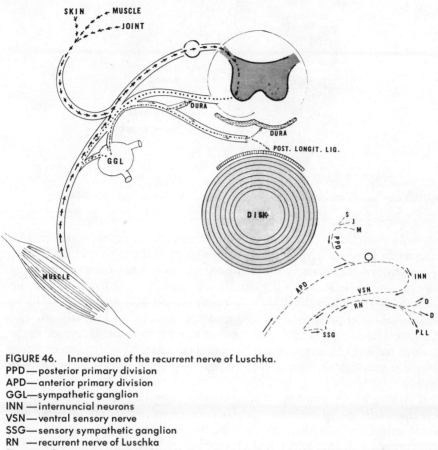

FIGURE 46. Innervation of the recurrent nerve of Luschka.
PPD — posterior primary division
APD — anterior primary division
GGL — sympathetic ganglion
INN — internuncial neurons
VSN — ventral sensory nerve
SSG — sensory sympathetic ganglion
RN — recurrent nerve of Luschka
D   — to dura
PLL — posterior longitudinal ligament

ultimately interpreted as pain from these sites of painful stimuli, form the basis of clinical evaluation of the patient complaining of low back pain.

STATIC AND KINETIC LOW BACK PAIN

Low back discomfort may be categorized as being either static or kinetic. The erect nonmoving spine (static) is capable of being painful and implies abnormal posture as the etiologic factor if congenital or organic pathology is not evident. The kinetic concept of low back pain implies faulty neuromusculoskeletal function.

The normal static spine requires a physiologic lumbar lordosis with the entire erect vertebral column in alignment with the center of gravity. In the physiologic lordosis there is no weight bearing upon the posterior

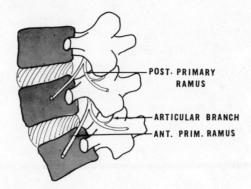

FIGURE 47. Innervation of posterior articulations (facets). Lateral oblique views.

articulations (facets); the intervertebral foramina are open and there is no abnormal pressure upon the posterior aspect of the intervertebral disk.

In abnormal posture with excessive lordosis there is weight bearing upon the posterior articulation. The foramina are narrowed and the posterior intervertebral disk is compressed and may protrude to encroach upon the posterior longitudinal ligament or laterally into the foramen to compress the dural sac of the nerve root. If any of these tissues have not been previously irritated or inflamed by trauma or stress, no pain may result, but, if the tissues have been sensitized, pain may result.

It has been demonstrated that merely maintaining a forward flexed erect posture of 30 degrees markedly increases intradiskal pressure and demands excessive muscular effort; therefore, this prolonged daily posture may be an incriminating factor in low back pain.

In faulty kinetics with resultant pain several aspects must be considered: (1) the tissues, ligaments or muscles may be restricted in their elasticity, thus limiting full flexibility; (2) the posterior thigh and leg muscles may be restrictive, thus preventing full pelvic rotation and causing the lumbar spine to assume a greater burden; (3) the lumbar-pelvic rhythm may be violated, especially in the process of reextension from the fully flexed position. This violation may be evident in the following manners.

a. Deceleration during forward flexion may be erratic and cause ligamentous and muscular strain.

b. Faulty reextension may have the patient reassume his lordosis before the derotation of his pelvis, thus causing all the stresses of the static lordosis with the addition of excessive muscular effort demanded by the eccentric position of the upper two thirds of the body.

c. Violation of the facet planes of movement may cause the body to reextend into lordosis while the spine is still abnormally rotated. In the deviation the facets are asymmetrically aligned and can be impinged on their counterpart, or the disk annular fibers can be exposed to excessive rotational torsional stresses with a fiber tearing.

It may be concluded from the concept expressed that faulty training, poor conditioning, poor musculature, and inflexible connective tissue play a large part in both kinetic and static low back pain. Fatigue, anxiety, anger, and distraction can also play a vital role in impairing proper neuromuscular coordination so vital to proper body mechanics.

The erector spinae musculature, which must become active and operational in the forward flexion as "decelerating" muscles until the entire body has been flexed to the point of the patient's hand touching the floor, requires that the muscles perform in an eccentric manner or so-called negative work. This is in contradiction to the concentric or positive work of reextension during which the muscles shorten.

In eccentric or negative work, during which the muscles lengthen during their contraction, each muscle fiber produces more tension than when it shortens during contraction. Oxygen cost of negative work is less than that required for the same amount of positive work although the same tension is produced. Therefore, energy expenditure does not differ much in concentric or eccentric shortening or lengthening of the musculature. Eccentric or negative work is also under voluntary control and requires a smooth neuromuscular control mechanism which is a matter of training as well as conditioning. To be smooth, no extraneous stresses must be imposed upon the neuromuscular mechanism. Extraneous mechanisms such as haste, anxiety, fatigue, or distraction can interrupt a smooth lengthening eccentric muscle contraction. An erratic, irregular descent of forward flexion may result if any of these factors are present. The muscles, joint capsule, ligaments, and intervertebral disks are subjected to erratic stressful forces.

Upon reaching full flexion with the person's fingertips touching the floor and the knees not flexed, the spine again becomes supported by the ligamentous structure with the muscles no longer in the state of contraction. Reextension to the erect posture requires the muscles' contracting in a positive or concentric manner of shortening. As they shorten, lumbar lordosis is regained and pelvic rotation returns to normal lumbosacral angulation. This motion, equally dependent on neuromuscular coordination, is also a product of training and competent musculature. Erratic contraction or shortening can result from improper training, faulty mechanism, distraction, fatigue, and impatience. All of these neuromuscular mechanisms can participate in the faulty neuromuscular skeletal mechanism of the lumbar-pelvic rhythm.

It may be concluded from this that faulty training, poor conditioning, poor musculature, inadequate flexibility, fatigue, anxiety, or distraction may result in a painful resultant misuse of the flexion/extension capabilities of the human spine.

Diagnosis

The history is invaluable in ascertaining whether the patient is suffering from static or kinetic low back abnormality. The history will indicate if the patient's symptomatology occurs from prolonged standing or sitting, and a description of the posture during this activity can be elicited. It is also possible by examination to evaluate faulty posture, excessive lordosis, or faulty occupational daily activities. Aggravation or initiation of the low back symptomatology by hyperextending the lumbosacral spine with reproduction of the identical low back pain ascertains that the lordotic posture is condusive to the symptomatology. The history may reveal the faulty mechanisms of flexion and reextension as the patient relates a story of improper bending and lifting and simultaneous twisting. All may have been present during the activity that resulted in low back pain. The mental state, environmental factors, distraction, or anxiety in patients can be elicited by careful history. Clarification of exact faulty kinetic mechanisms by examination may not be as easy as determining the static postural component, but watching the patient bend, stoop, squat, and lift, as well as sit, will frequently indicate the faulty mechanisms used by the patient. Rarely is laboratory work of any great significance other than to rule out the presence of metabolic, infectious, inflammatory, or malignant changes in the vertebral column or through chemical tests to substantiate the presence of organic pathology.

Treatment

Improvement of posture is the most important aspect of treatment of the static back. This requires realignment of all the physiologic curves to conform to the center of gravity. Strong abdominal muscles, which anteriorly lift the pelvis, and strong muscles of the buttocks, which lower the posterior aspect of the pelvis, tend to change the lumbosacral angle by tilting the pelvis or derotating the lordotic curve. Increased strength and endurance of the anterior abdominal musculature is of great importance but more important is the kinesthetic appreciation of the resultant pelvic rotation. Without the kinesthetic appreciation of pelvic tilting, good abdominal musculature will never be utilized even though present.

Rotation of the pelvis can be taught the patient by a two-stage exercise. Have the patient lie in the supine position on the floor or on a bed with knees and hips flexed and the lumbar spine pressed to the floor. The patient's hand may be superimposed between the lumbar curve and the floor to demonstrate the movement of the lumbar spine. This position is held while the pelvis is slowly elevated from the floor *without* elevating the lumbar spine from contact with the floor (Fig. 48).

This concept, once learned, can be done repeatedly with decreasing amount of effort. As the flattened lumbar curve becomes achievable, the hips and knees can be fully extended until both legs are fully extended and the pelvis still tilted to decrease the lumbar curve. This exercise and pelvic movement must ultimately be done in the erect posture. A wall, against which the exercise is performed, is substituted for the floor.

The lower back must be significantly flexible to achieve a full flatback posture. An exercise of knee-to-chest, knee-to-chest, and both knees-to-chest may be instituted (Fig. 49). In this exercise the patient holds his/her knee in his/her hands. This assures that the leg is used as a lever arm to flex the low back and not to hyperflex the knee, which results in stress on the knee without influencing the low back flexibility. It is desirable that the abdominal muscles be simultaneously contracted as the knees approach the chest during this exercise.

Forceful rhythmic exercise may be detrimental. Acute stretching of muscle spindles causes reflex contraction. After forcefully stretching

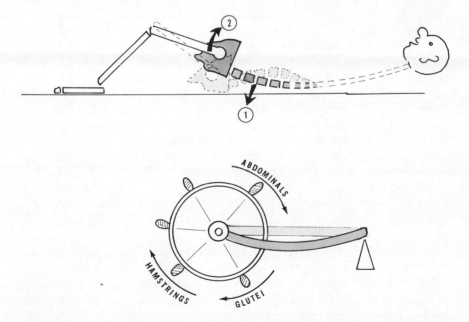

FIGURE 48. Concept of pelvic tilting. *Above, 1*, Flattening of lumbar curve against the floor, table, or bed; 2, Gentle rotation and elevation of pelvis; *Below*, Musculature involved in pelvic tilting.

71

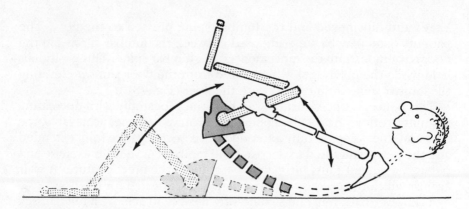

FIGURE 49.   Gentle low back stretching exercises using the femurs for leverage.

extensor muscles they may become further contracted. Gentle stretching and maintaining the acquired stretch position for brief periods is more physiologic.

A sit-up exercise is effective in strengthening the abdominal muscles, but it must be specifically prescribed and supervised. A sit-up from the full supine position may hyperextend the lumbar spine during the first phase of flexion. As there is simultaneous neck flexion, cervical strain may occur. Holding of the breath while doing this exercise may cause a Valsalva effect. All of these factors are undesirable. Therefore, the exercise program is best done in a different sequence. Beginning from the fully-flexed position, the body is gradually and progressively lowered to the floor (Fig. 50). The time required to achieve a full sit-up may be weeks with gradual progression being carefully supervised.

With hands near but not touching the knees during this exercise the patient is always assured that he can have manual assistance to resume the full flexed position if the abdominals become fatigued or painful.

Since the muscles that keep the pelvis tilted must function isometrically instead of isotonically, the muscles must be exercised for endurance as well as strength. A partial flexed position must be assumed and held for increasing durations of time to accomplish this (Fig. 51). The amount of resistance imposed upon the abdominals can be increased by changing the angles of the arms, which are positioned with the hands behind the head.

As a result of spinal segmental reciprocal innervation, contraction of the flexors reciprocally relax the extensors. Therefore, when erector spinae muscles contract to splint a painful lumbosacral spine, contraction of the abdominal muscles reciprocally relax these muscles and permit the back to flex to a greater degree.

Sustained isometric contraction of the abdominal musculature gives a proprioceptive sensation to the abdominals, which is helpful in sustaining

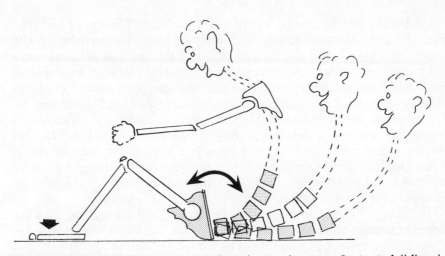

FIGURE 50. Abdominal flexion exercises from shortened position. Begins in full flexed position. The body is gradually lowered, the position sustained, and returned to upright position permitting full flexion gradually from supine position. Arms are held near knees to prevent excessive extension.

a tilted pelvis in the upright stance. It is important also to modify the faulty sitting and standing activities of the individual in order to assure that the pelvis remains tilted and the lordosis minimal throughout activities of daily living.

If faulty kinetic flexion reextension has been diagnosed as the cause of low back difficulty it is important to teach the patient how to bend, stoop, squat, and lift. The lumbar lordosis must always be decreased and any object lifted held close to the body. Bending and lifting must be performed avoiding any lateral flexion or rotation in flexion and most

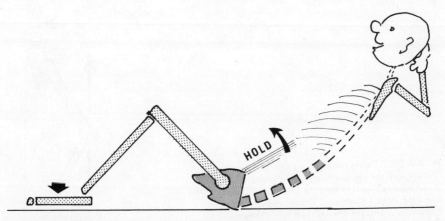

FIGURE 51. A partial flexed position used to strengthen pelvic muscles isometrically.

73

emphatically in reextension to the erect position. The knees and hips must be bent simultaneously with the bending, thus permitting the pelvis to remain tilted and tucked in under the spine as flexion occurs. This requires strong quadriceps muscles; therefore an exercise to strengthen the quadriceps must be undertaken along with all exercises relative to the back. Such an exercise very simply may imply flattening the back against the wall and slowly sliding down and back, sustaining a holding posture of semiflexion during the exercise series (Fig. 52). This causes simultaneous quadriceps exercises with a flat lumbar curve. Practicing spinal flexion/extension in the process of lifting eventually assures this becoming an automatic mechanism and one that will be violated only when distraction, fatigue, or haste is imposed.

In lifting (Fig. 53), an object is placed on the floor and the feet are brought parallel and directly facing the object. Both feet are advanced as close to the object as possible. The knees are flexed and slowly and smoothly the entire body is flexed to pick up the object. The last step is to

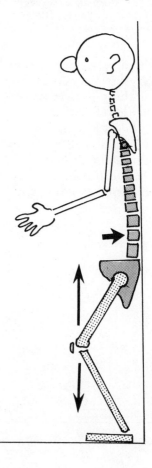

FIGURE 52. Pelvic tilting in upright position by flattening back against wall and sliding up and down against wall to initiate concept of proper bending and squatting. Strengthening quadriceps musculature are utilized simultaneously in this maneuver.

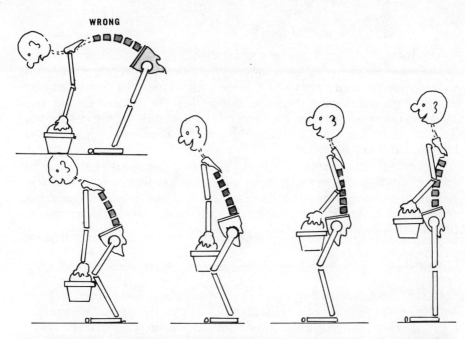

WRONG

FIGURE 53. Correct and incorrect method of lifting. *Above, Incorrect; Below,* Correct. Bending with simultaneous bending of the knees; the object held close to body. The entire body flexes as object is picked up. The spine re-extends to upright position with simultaneous extending of hips and knees.

reextend the spine to the erect position and simultaneously extend the hips and knees. The best method of training is to repeat this activity until it becomes automatic.

The patient must always be made aware that distraction, fatigue, impatience, anxiety, tension, or anger may violate the best trained pattern and cause the patient to assume either a faulty posture or faulty flexion and reextension. It is from this momentary violation that the back may become injured.

It is evident that the average mechanical low back discomfort can best be avoided by postural training and lifting-bending instruction and practice. The exercises needed for flexibility and muscular strength and endurance are only a portion of the daily therapeutic activities.

ACUTE PAINFUL EPISODES OF LOW BACK DISCOMFORT

Frequently an acute discomfort of the lower back does not permit adequate examination and only a history can be elicited. Treatment must precede accurate diagnosis. Absence or presence of neurologic abnormality must be ascertained early and this can be done even though the patient is in severe pain.

The patient who presents with a history of low back pain with referral of the pain into either or both lower extremities is currently considered to be suffering from diskogenic disease with sciatica. The onset may be acute or insidious with the pain localized in the lumbar region and radiating into the buttocks, the posterior or lateral thigh, the calf, or the ankle region. Onset may be related to a readily recalled incident or may be innocuous as to its cause.

As a rule there is gravitational influence upon the pain with the patient more uncomfortable in the sitting than in the standing or lying position. Bending, stooping, or lifting is avoided because of pain. Coughing, sneezing, or defecation often aggravates the pain. The reclining position frequently is beneficial.

Examination

In the moderate to severe low back pain the patient reveals (1) an antalgic spine, (2) functional scoliosis, (3) usually a positive straight leg raising test, and (4) subjective or objective neurologic deficit.

The antalgic posture is manifested as a reversal lumbar lordosis with the lumbar spine slightly kyphotic or erect. This posture is assumed to be due to the protective spasm of the multifid muscles.

A functional scoliosis may result in which the patient is scoliotic in the erect posture but regains his straight vertebral column in the prone position (Fig. 54).

The neurologic examination may be performed very simply by testing the deep tendon reflexes; superficial muscle testing of all the myotomes can be performed by resistance of the big toe extensor (L_5), anterior tibialis (L_5), and quadriceps (L_3 and L_4). The first and second sacral vertebrae, which innervate the gastrocnemius-soleus group, are difficult to examine in a bedridden patient. Only in the standing position rising up on the toes of one foot can weakness and fatigue be elicited, thus implying S_1 root involvement. The dermatomes can be elicited from the patient's history and confirmed by objective scratch, pinprick, or touch pattern testing (Fig. 55).

Straight leg raising is a test that merits consideration. This test, commonly called the Lasegue test and now simplified to straight leg raising test, is termed positive or negative with an implication of pathologic factors when termed positive. The straight leg raising test was described originally by Forst[3] and ultimately by Lasegue.[4] Both Forst and Lasegue attributed a positive straight leg raising test to be caused by muscle spasm. Later, when it was noted that the lower lumbar roots moved within the intervertebral foramen during straight leg raising,

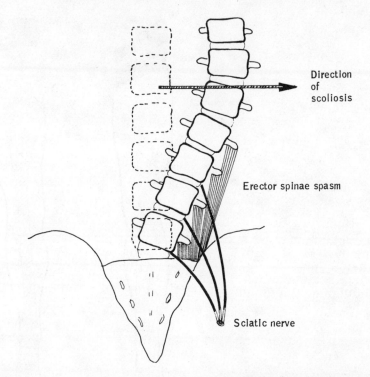

Direction of scoliosis

Erector spinae spasm

Sciatic nerve

FIGURE 54. Muscular component of functional scoliosis. The deep interspinous muscles including the multifid muscle act unilaterally to pull the spine and create a functional scoliosis. Spasm of the muscle, considered to splint and thus be protective, is considered to be involved.

sciatica was attributed to tension upon the inflamed foot. This observation ultimately became the basis for considering a positive straight leg raising test as confirmatory evidence of nerve root pressure from a herniated lumbar disk (Fig. 56).

There are many modifications for performing the test. Originally the test was described by Forst and Lasegue as flexion of the hip to 90 degree and then extending the leg upon the thigh. It is now performed by elevating the entire extended leg with the patient supine; however, the original Forst-Lasegue method is utilized when the patient is seated and the leg extended to the horizontal position. Insofar as nerve root movement within the intervertebral foramen is not considered to occur within the first 30 degrees of straight leg raising, a positive test is claimed after 30 degrees of elevation.

Questions have arisen regarding pain along the sciatic nerve distribution caused by pressure upon the nerve. It is clinically ascertained that compression of normal peripheral nerves or nerve roots causes a loss

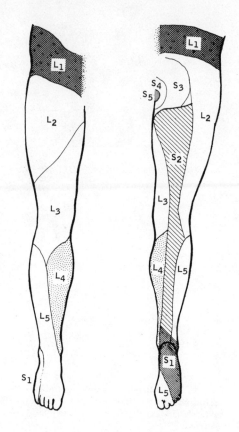

FIGURE 55. Dermatome distribution of lower extremities.

of function but no pain. Rapid compression causes paresthesias such as tingling and possibly brief transient pain.

In patients who were found to have inflamed nerves, merely touching these nerves caused pain. Lindahl[5] placed a cannula in the epidural space and rapidly inflated the cannula. This caused pain only in previously inflamed nerves. Granit and coworkers[6] postulated that hyperalgesic nerves were nerves which lost some insulation (demyelinated) and, thus, were susceptible to subsequent touch or pressure. McNab[7] demonstrated that, under direct observation, mere compression of an inflamed root with a small surgical instrument could cause pain in the root distribution.

Recently it has been shown that interruption of the posterior primary ramus can diminish or eliminate sciatic nerve root distribution (an anterior primary nerve root) and allow painfree straight leg raising (Fig. 57).

Patients, who have had diskectomies but suffered a recurrence of radicular pain, have had residual numbness and recurrence of radicular pain, or actually have had *no* benefit from surgery in respect to relief of

78

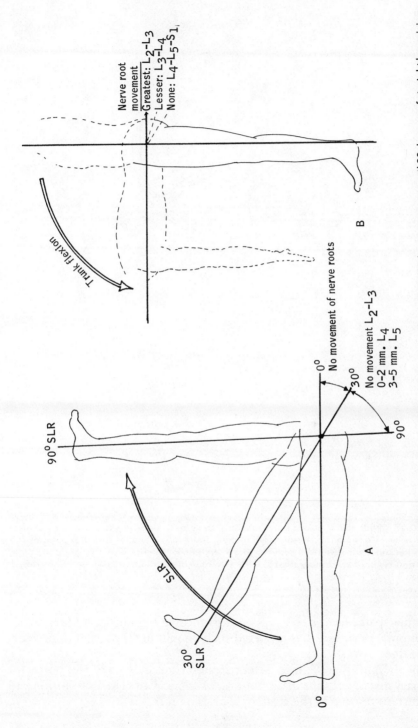

FIGURE 56. Straight leg raising with motion of the nerve root at the intervertebral foramen. *A,* With no movement until 30 degrees of straight leg raising, following which movement of up to 2 mm. at L4 root and 3 to 5 mm. at L5 root. *B,* Movement of nerve roots and foramina on forward trunk flexion as depicted.

Within figure A:

90° SLR

30° SLR

SLR

0°

30°

90°

0°

No movement of nerve roots

No movement L2–L3
0–2 mm. L4
3–5 mm. L5

A

Within figure B:

Trunk Flexion

Nerve root movement
Greatest: L2–L3
Lesser: L3–L4
None: L4–L5–S1

B

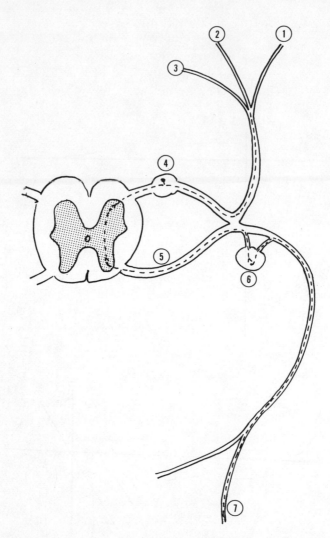

FIGURE 57. Radiation of sciatic nerve into anterior primary division with noxious irritation at 1, posterior skin region; 2, erector spinae musculature; 3, posterior articulations (facets). Radiation proceeds down the sensory root 4, and down the anterior primary division 5, gathering sympathetic innervation 6, with dermatomal distribution down anterior primary division 7.

radicular pain, have been reported to receive benefit from dorsal root rhizotomy. Bertrand[8] has reported such benefit in 40 percent of patients so treated.

Shealy[9] and Rees[10] have demonstrated that division of the posterior primary division results in diminution of sciatic root pain distribution and a return to normal in the straight leg raising test.

80

Manipulation of the posterior facets, which are innervated by the articular branch of the posterior primary division, has also been claimed to relieve sciatica and permit normal straight leg raising.

Injection of irritating solution within the posterior facet joints has produced dermatomal type of root distribution, the exact site of injection into the joint being verified by a facet arthrogram preceding the injection of irritating solution. Relief can follow this injection by injecting a local anesthetic into the facet joint.

Tender areas within the musculature, felt as nodules, are commonly termed trigger points and may be injected with a local anesthetic with resultant relief of the local back pain and the area of referral down the leg (see Fig. 17).

From the above discussion and all the factors implicated, the diagnosis of *lumbar diskogenic sciatica* raises many questions of pain causation. The "natural history" of this disease syndrome with numerous spontaneous recoveries regardless of the type of treatment (sciatica without detectable disk herniation)[11] and the frequent failure of surgical diskectomy suggests that actual disk protrusion may be incidental, only partially related, or may actually be a complication rather than the primary condition. If the latter is true, what the primary causative condition is remains for future edification.

Treatment

Complete bedrest with elimination of gravity and elimination of pain is the keystone of treatment. Bedrest must imply *rest* in every sense of the word. The position of choice is the position of greatest comfort to the patient, whether it be full extension on a firm bed or semi-flexion.

Flexion position in bed, erroneously termed semi-Fowler position, indicates flexion of the hips and knees with a slight sitting-up position. This is easily achieved in the hospital by mechanically posturing the bed. It is more difficult to achieve this position at home.

To assume a semi-flexed position at home, placement of pillows under the knees is difficult to maintain and unpredictable in determining the degree of flexion. One or more flat, square pillows, commonly found on many sofas, may be placed under the patient's knees and preferably placed under the bedsheets to remain more securely on the bed. (Fig. 58). The average head pillow completes the semi-total flexed position.

Rest also implies emotional and mental relaxation. Pain must be relieved by proper medication. Restriction of activities must be very specifically outlined by the physician or therapist. This relates to meals, bath privileges, daily activities, home activities, and so forth. No activity should be left to the discretion of the patient. All restrictions must be facilitated in a realistic manner. Emotional relaxation may require

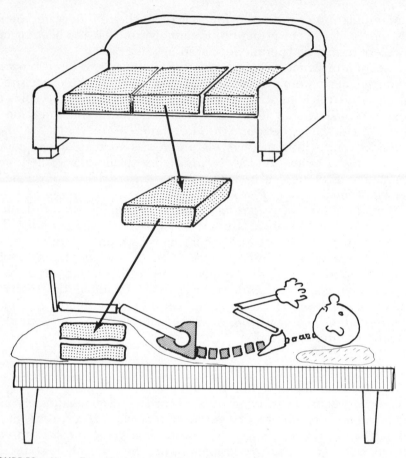

FIGURE 58. Home flexed bed posture utilizing square sofa pillows inserted under bedsheets to facilitate flexed hip and knee posture in lumbar diskogenic disease.

tranquilizing, which is safe so long as the patient does not need to ambulate or function to any significant degree.

MANIPULATION. Manipulation has been advocated by many as effective treatment in the acute phase of low back pain that exhibits the antalgic posture, functional scoliosis, and subjective sciatic radiation. Insofar as most cases recover spontaneously by simple bedrest of relatively short duration, treatment by manipulation is usually not indicated.[12]

Intramuscular injection of an analgesic agent is often effective in patients who exhibit evidence of paraspinous spasm with tenderness of the erector spinae muscles. Injections of 3 to 5 cc. of lidocaine or a similar anesthetic agent can be injected into the area of maximum tenderness $1\frac{1}{2}$ to 2 inches lateral to the spinous process.

Frequently the lumbosacral fascia (Fig. 59) is a source of tenderness with deep pain and a concurrent segmental scoliosis. An injection of local anesthetic in this region is immediately beneficial in acute lumbosacral strain.

Tenderness of the supraspinal ligament, immediately overlying the affected disk or posterior articular structures, has been postulated. Also, an injury of a severe degree resulting in low back pain may disrupt the integrity of the supraspinal ligament and remain as a chronic source of pain and tenderness. Frequently, an injection of an anesthetic agent is both diagnostic and therapeutically beneficial.

HEAT AND ICE. The local use of heat may be of value for subjective relaxation and improvement. Application of this modality must, however, be possible with the minimal amount of movement of the patient and without altering the comfortable position assumed by that patient. An electric pad is controllable, easily applied, and usually readily available. Hot moist packs are more difficult to apply and to maintain at the correct temperature. Forms of heat such as diathermy and microwave are not readily available in the home situation.

Application of ice has been advocated by many and proven to be beneficial. Ice apparently decreases congestion, anesthetizes the sensory skin fibers, and reflexly relaxes the underlying muscles.

An ice applicator that can be administered by the patient, his therapist, or a member of the family can be simply made (Fig. 60). Instructions are to fill a paper cup with water, insert a tongue blade into the water, place

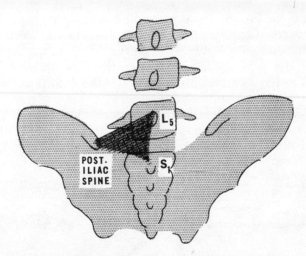

FIGURE 59. Lumbosacral fascia. The lumbosacral fascia is triangular: the borders being from the posterior iliac spine, the posterior superior spine of L₅ and the upper border of the sacrum. This is a frequent site of myofascial pain and an injection into this fascia frequently affords relief.

FIGURE 60. Practical home ice application. A paper cup filled with water into which is inserted a tongue blade is placed in the freezer. When frozen, the paper cup is removed.

the cup in a freezer until solid, and remove the paper cup. Gently rub the applicator over the area to be iced until the site assumes a slightly pink hue; discontinue icing. This can be repeated hourly.

The choice of modality, be it heat or ice, is a choice based on the experience of the therapist or the preference and benefit expressed by the patient.

TRACTION. Traction has been advocated for diskogenic disease for several decades. Various weights, methods, and angles of pull have been recommended but none have had full unqualified acceptance. The most commonly used lumbar traction has been pelvic traction with the pull applied to the pelvis or the legs. The most that can be said for this form of traction is that, properly applied, it decreases the lumbar lordosis and, thus, increases the foraminal opening and separates the posterior facets. Many even claim its only benefit has been the assurance that the patient remains in bed.

Traction applied to the leg as in Buck's traction does not decrease the lordosis but actually may increase the lordosis (Fig. 61). Because the hip joints are located anteriorly to the sacrum, pull on the femur causes rotation of the pelvis, which increases the lumbar lordosis. Buck's attachment to the lower leg also makes it difficult to fully examine the patient and it restricts bathroom privileges.

The standard pelvic belt traction has a mechanical disadvantage in that it does not take into consideration the friction component of the body weight in the bed or treatment table. The friction of the body creates a

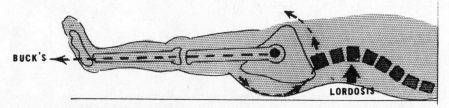

BUCK'S

LORDOSIS

FIGURE 61. Buck's traction. Because the line of pull is anterior to the lumbosacral joint, rotation of the pelvis can increase the lordosis.

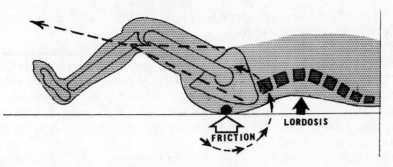

FIGURE 62. Pelvic belt traction: friction factor. The pelvic band traction acts to pull on the pelvis. The friction of the body upon the bed or table causes a pivot point about which the pelvis rotates. This specific rotation increases lordosis.

fulcrum about which the pelvis rotates and, thus, causes rotation that increases lumbar lordosis (Fig. 62). The friction also neutralizes much of the traction force.

By supporting and elevating the legs sufficiently the pelvis is elevated from the bed. The weight of the body above the pelvis decreases the lordosis and, thus, accomplishes part of the benefit of traction. Traction can be applied simultaneously (Fig. 63).

Pelvic traction with a single central posterior strap attached to the pelvic belt at the level of the sacrum rotates the pelvis mechanically and

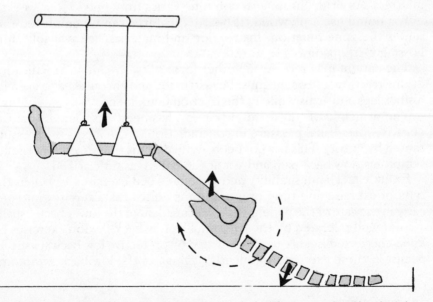

FIGURE 63. Elevation pelvic traction. By suspending the legs from an overhead bar the pelvis is elevated and this decreases lumbar lordosis.

decreases the friction component (Fig. 64). When used with a divided table, friction is further reduced.

Recently a promising form of traction has been employed at the Sister Kenny Institute. This utilizes the idea that the weight of the body below the third lumbar vertebra is estimated as 30 percent of the total body weight (Figs. 65 and 66). The traction belt is applied to the chest area by a fitted harness that firmly secures the rib cage. The harness is designed to have the lowest strap under the rib cage and the uppermost strap securing the upper rib cage.

The body is inclined with the pull gradually approaching full upright position—90 degrees to floor level. At this ultimate angulation the full lower body weight applies the traction (Fig. 67).

Traction is applied at first in a slight angle of approximately 30 degrees for increasing periods of time. At first, four hours is usually well tolerated. On a carefully controlled and recorded schedule the angle of pull gradually increases to 90 degrees and for longer periods of time.

The harness is removed periodically during the day for skin care with body lotion or powder and for rest. During rest periods the patient is prevented from sitting, standing, or walking except for bathroom privileges.

In the hospital setting, this type of traction can be installed into the electrically-operated stryker bed. The patient controls the angulation and the time. A footplate is placed several inches below the patient's foot in case the chest harness slips, thus preventing the patient's sliding off the mattress. As a rule, hospitalization time varies from one to four weeks.

For home use a plywood tiltboard apparatus can be employed that serves the same function. Instruction and materials are available from Lossing Orthopedics.*

The patient may receive any other form of treatment and medication during traction. He or she must be instructed and encouraged to exercise his/her legs and arms while in the traction equipment.

The principle of this form of traction has been postulated to create negative intradiskal pressure in contradiction to the increased pressure caused by gravity. This has not been verified nor is the reason for relief of symptoms (low back pain and sciatic radiculitis) as yet clarified.

EXERCISE. The desirability and benefit of bed exercises as well as the type of exercise and time of initiation is equivocal. Exercises initiated early in the care of the patient with an acute pain of the lower back usually are not well tolerated by the patient. The low back flexion exercises by knee-chest exercises are usually negated by protective low back spasm. To persist in these exercises frequently aggravates the low back symptoms.

---

*Lossing Orthopedics, 2217 Nicollet Ave, Minneapolis, Minnesota 55401.

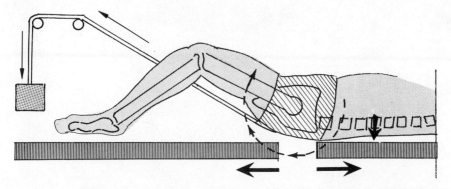

FIGURE 64.   Single strap pelvic traction. With a central posterior strap attached to the pelvic band the traction pull rotates the pelvis, maximizes friction, and decreases lordosis. A split table, when elongated, further decreases friction.

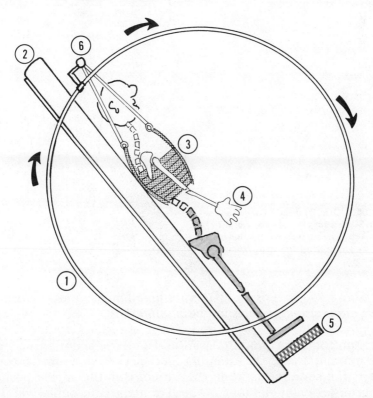

FIGURE 65.   Gravity lumbar traction (Sister Kenny Institute). 1, Lumbar traction applied in a hospital setting with a circular bed; 2, Mattress with bedboard within Stryker circ-o-lectric bed; 3, Chest harness with lower straps under the rib cage and upper straps firmly grasping the rib cage; 4, Bed manually controlled by patient; 5, Footplate several inches below patient's feet for security; 6, Snap ring attached to bed frame.

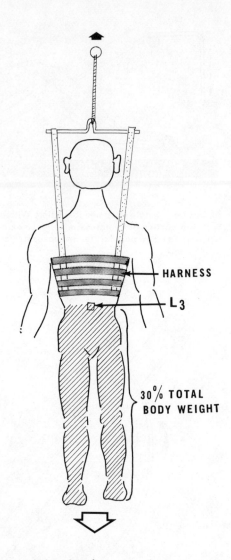

HARNESS

L₃

30% TOTAL BODY WEIGHT

FIGURE 66. Gravity lumbar traction. Based on 30 percent of total body weight below the third lumbar (L₃) level traction is applied by supporting the body from a chest harness.

Pelvic tilting also is resisted by constraining lumbar muscle spasm. A Valsalva effect also aggravates low back symptoms.

It is possible, however, to take advantage of reciprocal relaxation via agonist-antagonist reciprocal relaxation. Isometric abdominal contraction reciprocally relaxes the extensors. Since neck flexion simultaneously contracts the abdominals when the patient is in the supine position, gentle neck flexion by elevating the head from the table or bed will cause simultaneous abdominal contraction. There will be simultaneous extensor relaxation in both the cervical region and the lumbar region.

Abdominal isometric contractions can be initiated gradually. This exercise will (1) accomplish extensor relaxation, (2) maintain abdominal

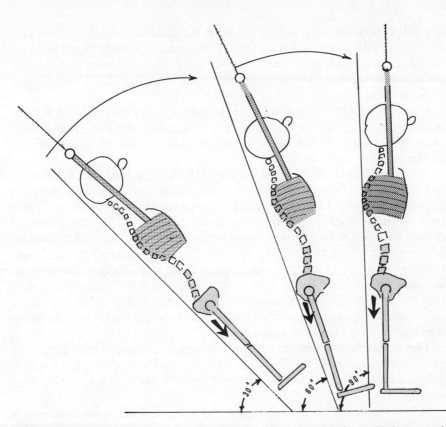

FIGURE 67. Gravity lumbar traction. As the bed (traction) angle goes from a 30 degree angle to full upright (90 degree) the amount of traction increases. The duration of traction can be determined.

tone, and (3) retain the concept of abdominal contraction to ultimately use in pelvic tilting.

Gentle pelvic tilting exercises previously described can be started gently early in the complete bedrest phase of treatment. This exercise is initially performed with the knees and hips flexed and gradually extended.

RESUMING ERECT POSTURE. After a period of bedrest, which may vary from five to fifteen days it is appropriate to begin sitting and then standing. Sitting must be performed by gently rolling to one side simultaneously flexing the hips toward the abdomen and chest. While doing this, the abdominals must be kept firm. This is best accomplished by raising the head, shoulders, and arms gently until the abdominals are firm and rolling over to the side as the hips and knees become gradually flexed. The arms are then used by pulling on the edge of the mattress and rising gradually to the flexed arm upon the elbow. Therefore, sitting to the full position is accomplished with simultaneous arm motion and assistance with sustained abdominal contracture. Returning to the bed position must be done in the exact reversal of this maneuver. Sitting up in bed directly from the supine position to the seated position is definitely to be avoided.

Upon standing, the patient must bring his legs as directly under the body as possible. With the abdominals contracted, the patient gradually stands erect with the use of the arms, *always maintaining the abdominals contracted and the low back "flat."*

BACK SUPPORT. If the acute injury of the low back has been severe or is a recurrence of numerous previous similar episodes, back support should be provided for the initial period of sitting and standing. This is best accomplished by a lumbosacral corset (Fig. 68) in which several principles of corseting or bracing must be observed. The corset must be of adequate length to extend posteriorly from the sacrum to the lower thoracic area. As a rule, this is a distance of 10 to 15 inches. Flat stays to decrease the amount of lordosis may be inserted into the corset; usually two will suffice as they are essentially a guidance to the flatback posture. An anterior abdominal uplift support with the abdominal pressure being superpubic is definitely indicated.

This corset will substitute for the weakened abdominal muscles if they exist. It will insure the back's being kept in a moderate decreased lordotic posture and will minimize any flexion or rotation of the spine in most activities.

When the patient resumes the erect posture and becomes ambulatory, it is mandatory that thorough instruction ensue in which proper posture and proper bodily function be stressed. Proper posture requires alignment of the entire body in relationship to the center of gravity. The postural curves must be minimized.

Because the entire superincumbent vertebral column is dependent upon the angulation of the sacral base this lumbosacral angle must be minimized. Tilting of the pelvis actively accomplishes this and simultaneously decreases lumbar lordosis.

The most stable position of the erect spinal column is when the line of vertebral forces passes through the sacrum and the least stable is when the line passes in front of the sacrum as it does in excessive lordosis. As shown radiologically by Jonck,[13] heavy loads carried upon the head causes compensatory movement at the lumbosacral spine with a change in the lumbosacral angle. Therefore, a technique for improving posture and decreasing the lumbar lordosis is the application of increasingly heavy weights upon the patient's head in order to train the patient both in standing and walking, and even in sitting (Figs. 69 and 70).

INJECTIONS. At the end of two weeks, if symptoms persist such as posture straight leg raising, functional scoliosis, or if there has been no

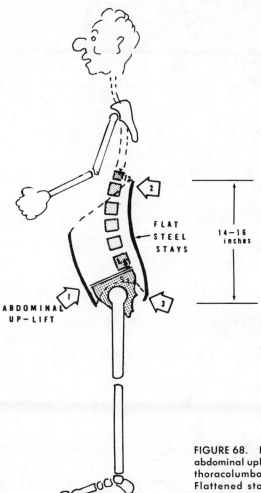

FIGURE 68. Proper lumbar corseting with *1*, Firm abdominal uplift support. Back point of contact at 2, thoracolumbar junction, and 3, over the sacrum. Flattened stays decrease lordosis and restrict activities requiring flexion and extension.

progression in minimal neurologic defect signs despite the fact that bedrest has been adequate in that it has been maintained, activities minimized, and medication for pain adequate, then intradural or extradural injections of steroids should be considered.

There is negative pressure within the epidural space so, as a needle-syringe penetrates the space, the plunger is pulled into the syringe barrel. If the needle inadvertently penetrates further and enters the dura the spinal fluid there is under pressure and the outflowing spinal fluid forces the plunger out of the syringe. These actions of the syringe are guidelines which indicate to the physician whether the injection is epidural or intradural.

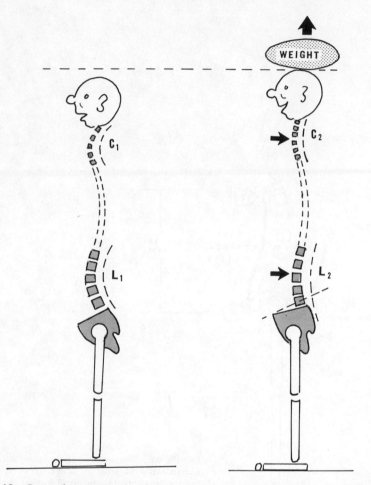

FIGURE 69. Postural correction. By placing a weight upon the head and elevating it the posture is improved in that the lumbar and cervical curves are decreased and the body approaches the center of gravity.

To perform an epidural block (injection of an anesthetic) the patient preferably sits in a forward flexed position with the back to the physician. A 17 gauge Touhy thin-walled needle with a Huber point is recommended. The space between the second and third lumbar disks (L_2 and L_3) or the third and fourth lumbar disks (L_3 and L_4) is located. After penetrating the skin a depth of approximately 3 cm. resistance is met from the interspinal ligament (Fig. 71). At this point the stylette of the needle is removed and the syringe is connected. The plunger of the syringe should be inserted halfway into the barrel to demonstrate whether it will be pulled further in (from negative pressure of the

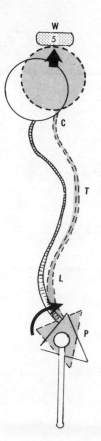

FIGURE 70. Postural change from distraction. The weight placed upon the head causes the pelvis to rotate and then decreases the lumbar lordosis. The superincumbent curves decrease and posture is improved.

W—weight
C—cervical vertebrae
T—thoracic vertebrae
L—lumbar vertebrae
P—pelvis

epidural space) or forced out (from positive pressure of the spinal fluid of the intradural space).

Once it is ascertained that the epidural space has been entered air is injected to distend the sac. The injectable material can then be injected directly through the needle or a vinyl or Teflon catheter can be passed through the needle into the epidural space. Ten to thirty cc. of sterile isotonic saline diluting procaine to make a 0.25 percent solution of procaine with or without a steroid added to the solution can be injected directly into the epidural space.

Lumbar paravertebral sympathetic blocks, long advocated to relieve radiating pain considered to be the result of the sympathetic component of nerve root irritation, are very difficult to perform and difficult to evaluate, and so in most clinics they are being replaced by differential spinal (intradural) blocks.

To perform a successful differential spinal block the patient must be cooperative and able to intelligently communicate. Before the test the

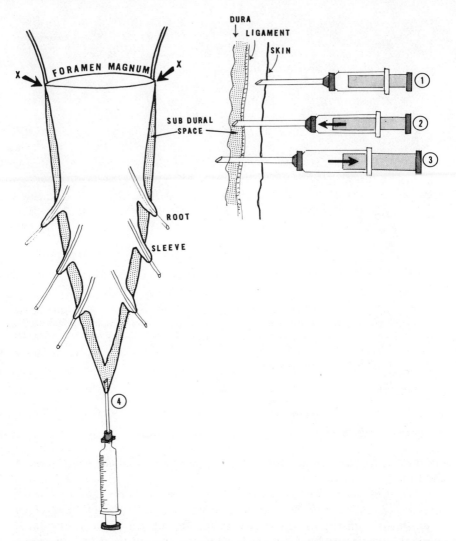

FIGURE 71. Dural sac with closure at the foramen magnum and the nerve root sleeves following the nerve roots through the intervertebral foramen. *Right 1*, Penetration of skin with spinal needle. *2*, Penetration of ligament into subdural space with the vacuum pulling the plunger into the syringe. *3*, Penetration of dura and into spinal canal with the spinal fluid ejecting the plunger from the syringe. The steps are the signs utilized by the physician performing an epidural or intradural injection to determine the depth before injection. *4*, Site of needle entrance into epidural sac: caudal injection.

patient must be carefully tested by pinprick and light touch sensory mapping, having a graded muscle test, and have the sympathetic system evaluated by skin temperature testing.

The spinal block, intradural injection, is performed at the level of the clinical site of complaint—usually the third to fourth or fourth to fifth

94

lumbar disk. Initially 5 ml. of normal saline is injected. A psychogenic basis for the pain is considered if the patient claims relief following injection. The needle having been left in place, the next injection is of 5 ml. of 0.2 percent procaine in isotonic saline. This dilution blocks sympathetic nerves and will give relief if pain is mediated through the sympathetic system. A third injection of 5 ml. of 0.5 percent procaine will block the sensory fibers. During this test both the subject and his blood pressure benefit from the dilution and must be documented as well as the objective sensory testing.

Benefit from saline injection suggests a large psychogenic factor. Benefit from the dilute solution implicates the sympathetic nervous system as the site of pain. When the 0.5 percent procaine solution affords relief the cause of pain is nerve root pressure. Further studies such as myelography may be indicated.

Further differentiation of somatic mechanism versus sympathetic transmission can be established by selectively injecting the peripheral nerve at the site where they anatomically separate. This technique, relatively simple in the upper extremity, can be difficult in the lumbosacral area.

Nerve roots can be infiltrated selectively with an anesthetic agent to identify specifically the offending root or to verify that the radicular pain is evoked from root irritation at the foraminal level. The technique requires an x-ray image intensifier to locate the specific anatomic region. With the patient in a prone position a hemostat is placed upon the patient's back to localize the surface area of the underlying bony structure.

The fifth lumbar (L_5) root is infiltrated by inserting an 18 gauge spinal needle at a 45 degree angle to the sagittal plane, 6 cm. lateral to the midline, and 4 cm. cephalad to the transverse process of the fifth lumbar root as located under fluoroscopy (Fig. 72). The first sacral (S_1) root is reached by injection into the sacral foramen.

The posterior articular facets have previously been considered the site of pain in the low back with radicular pain of sciatic distribution. This area was ignored after demonstration of nucleus pulposus protrusion as the cause of sciatica.

Hirsch[14] reproduced pain in the back with radiation into the upper thigh by injecting hypertonic saline into the region of the facet joint. More recently, injection into the facet joint, as verified by performance of arthrography, has reproduced radiation of pain down the posterior lateral aspect of the leg. The arthrogram cannot be interpreted other than its verifying the needle is in the joint.

Injection of irritating hypertonic saline into the third and fourth lumbar (L_3, L_4) facet produces pain down the anterior aspect of the leg and the fifth lumbar to first sacral (L_5, S_1) disks, more towards the ischium

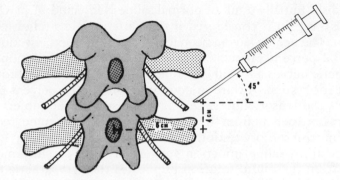

FIGURE 72. Technique of nerve block with injection being placed lateral to midpoint between two posterior spines at a distance approximately 6 cm. from midline and the needle being directed towards the midline at a 45 degree angle.

and the buttocks (Fig. 73). After injection of the L4 to L5 and the L5 to S1 facet joints there is marked activity of the hamstring muscles as noted on electromyographic studies. This could well explain limitation of straight leg raising (SLR).

The facet joint can then be blocked by an injection of a mixture of 1 percent Xylocaine and Depo-Medrol which, if successful, will give relief of pain and often permits full straight leg raising and normal flexion and reextension of the lumbar spine.

Facet injections are usually performed in a radiology laboratory under fluoroscopic control. The patient lies prone on the x-ray table with a pillow under the abdomen to flex the lumbar spine. By placing a towel clip on the back and then viewing it fluoroscopically, the exact location of the articular facet can be determined. An 18 or 20 gauge 3½ inch long needle is inserted approximately 1 inch lateral to the midline and directed down until bone is contacted. The superior facet can be identified fluoroscopically. By gentle movement the tip of the needle can be placed into the joint. As the joint is entered the operator can experience a "popping" sound or sensation and the patient usually experiences some local discomfort.

A metal grid may be constructed with wires spaced at 1 to 2 inches apart. This grid is taped on the patient's back and an x-ray picture will specify the exact intersection of the wires overlying the facet to be injected (Fig. 74).

If verification that the needle is in the joint is needed an arthrogram can be performed by injecting 3 to 4 cc. of Conray 60 solution through the inserted needle. Injection of hypertonic sterile saline solution into the joint can reproduce the patient's symptoms and verify this joint as the locus of pain production. Injection of an anesthetic agent (Lidocaine,

96

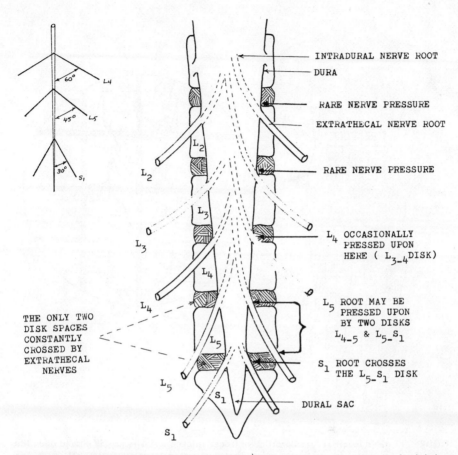

FIGURE 73. Relationship of spinal nerve roots and their dura to lumbar intervertebral disks. Site of nerve root compression is indicated.

Xylocaine, etc.) into the joint may relieve the patient's symptoms, thus further confirming the diagnosis as well as affording the patient relief. The injection of a steroid as well as an anesthetic is considered longer-acting in its relief.

Of the many nonsurgical methods for treating lumbar radiculopathy (sciatica), the use of corticosteroids have been widely advocated. This has been based upon the observation of marked inflammation and edema about the nerve root during surgical exposure. The frequent elevated spinal fluid protein in clinical herniated lumbar disk also favors an inflammatory process. Berg in 1953[15] correlated the decrease in the swelling of the lumbar roots with clinical improvement and stated, "It is not the disc protrusion as such but secondary pathognomonic changes which are responsible for the symptoms. These changes are probably inflammatory in nature." It is also well documented that patients with

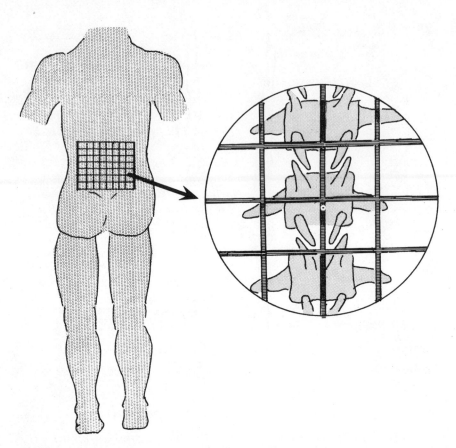

FIGURE 74. Grid localizer for interarticular (facet) injection. A wire grid is placed upon the back of the prone patient and then x-rayed. The exact wire intersection overlying the facet can be localized on the patient. Injection can be made without an image intensifier, thus decreasing patient exposure to irradiation.

herniated disks may have complete remission after conservative treatment without any change in myelographic defect.

In 1975 Green[16] described the use of dexamethasone in large doses for a period of one week as proving effective in patients having classic signs of herniated lumbar disk verified by myelography, electromyographic changes, and abnormal neurologic deficits. The dose advocated was 65 mgm. first day, 32 mgm. second day, 24 mgm. third day, 12 mgm. fourth day, and 8 mgm. the remaining four days. Patients who had recurrence after this series allegedly had milder symptoms and in the author's experience received good benefit from a second and smaller dose course. Dexamethasone administration by the intramuscular route was advocated but currently oral medication is proving effective.

Steroid therapy either intramuscularly, orally, intradurally, or extradurally presents another nonsurgical approach in treatment of the lumbar diskogenic syndrome that merits utilization.

SURGERY. There are patients who have intense radicular pain with positive Lasegue sign and minimal or no neurologic deficits attributable to entrapment of the nerve root in the lateral recess (Fig. 75). The entrapment is a stenosis of the intervertebral foramina resulting from encroachment by a hypertrophic superior articular facet. Degenerative arthritic changes of that joint usually account for the hypertrophy but occasionally congenital or developmental changes can account for the abnormality.

Myelography is usually not informative and measurement of the vertical and transverse diameter of the spinal canal are within normal limits. Exploratory surgery and superior facetectomy are indicated and beneficial.

PIRIFORM SYNDROME

Low back pain with sciatic radiation may be caused by entrapment of the sciatic nerve as it emerges under the piriform (piriformis) muscle. This is termed the piriform syndrome. There is a higher incidence in females (6:1) and no causative factors predominate.

On physical examination the lumbar spine has complete range of motion. A restricted straight leg raising may be elicited and this test is more probable when the leg is simultaneously rotated. Pain is more consistently elicited by resisting abduction and external rotation of the extended leg with the patient in a supine position. There frequently is localized deep tenderness over the sciatic notch. Abduction and external rotation of the hip may be weaker on the affected side. This may be tested by resisting the patient's ability to separate his/her legs while in the seated position.

The piriform muscle (Fig. 76), an external rotator of the hips, arises from inside the pelvis in the region of the sacrum and the sacroiliac joint. It passes laterally out of the sciatic notch to insert upon the greater trochanter of the femur. The belly of the muscle passes over the sciatic nerve. The sciatic nerve passes between *two* bellies of the muscle in 15

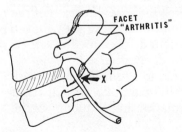

FACET
"ARTHRITIS"

X

FIGURE 75. Facet arthritis resulting from disk degeneration and articular change of the lumbosacral spine encroaches upon the intervertebral foramina causing low back pain from facet synovitis and nerve root entrapment at X with radicular pain.

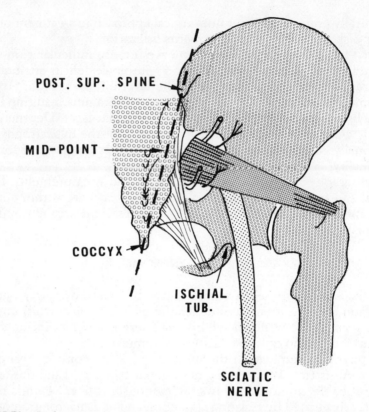

POST. SUP. SPINE

MID-POINT

COCCYX

ISCHIAL
TUB.

SCIATIC
NERVE

FIGURE 76. Piriform muscle. The sciatic nerve leaves the pelvis along with the piriform muscle through the greater sciatic foramen. The sciatic nerve emerges below the piriform midway between the ischial tuberosity and the greater trochanter. The emergence of the piriform muscle is midway between the posterosuperior iliac spine and the tip of the coccyx. These landmarks aid in determining the site of injection into the piriform muscle.

percent of people. The piriform muscle can be palpated by way of a rectal examination or pelvic examination.

Treatment requires direct local injection of the piriform by an anesthetic agent. There are numerous approaches. One way is vaginally or along the vaginal wall. However, it is simpler to insert a gloved finger rectally until the tender spot is located and then inserting a spinal needle through the buttocks through the sacral notch into the direction of the gloved finger. Since the patient is awake the injection cannot be given into the sciatic nerve. If the nerve is touched, the needle can be withdrawn a short distance and the anesthetic agent injected.

The piriform syndrome should be suspected in a female patient with normal lumbar spine motion, dyspareunia, and pain of sciatic distribution that is reproduced by resisting external rotation and abduction of the extended hip.

Although the mechanism of low back pain frequently can be anatomically and neurophysiologically verified and faulty posture and errant body utilization can be attributed as the cause of resultant pain, pain per se is an emotional response to sensory input. To deny pain as a psychologic response is unwarranted.

Secondary gains from pain located in the lumbosacral area have voluminous substantiation in the medical and psychologic literature and can only be briefly acknowledged here. Gratification from restrictions imposed by the low back pain and psychologic as well as monetary gains from incurred low back injury are legend. To deny their existence would be as inaccurate as to deny the existence of lumbar disk herniation. No patient having chronic recurrent or persistent disabling low back pain that impairs his social, personal, and occupational performance should be considered as thoroughly evaluated and competently treated until his psychologic function is adequately understood.

The sick person—here the patient with chronic low back pain—occupies a unique social position of being freed from daily activities and dependent upon others for his well-being. The person with severe low back pain is exempt from social responsibilities, cannot be expected to take care of himself, and is expected to seek medical advice and cooperate with these medical experts. Socially a person is "sick" when others identify and treat him as sick. Pain indicates abnormality in health and prevents a person from performing normal work activities or other social, even familial, functions.

Pain and disability that is rewarded tends to persist whereas unrewarded pain and "sickness" loses it benefit. This simplistically is the lesson of operant conditioning. The less certain or the poorer the prognosis of an illness or a disability the greater is the tendency to define a person as "sick." This label is magnified as recurrent failures of various treatment modalities are experienced.

Measurement of Pain

Measurement of pain and human reaction to pain remains the greatest challenge to clinical practice. Objectivity, as compared to subjectivity, is uppermost in the mind of the diagnostician and, failing to find an objective basis of pain production, becomes a source of frustration.

Chronic pain localized in the lumbosacral area undoubtedly has an emotional component and all patients presenting with these complaints must have a psychologic evaluation to determine any magnification of the pain claimed by the patient. Intervention, in many instances, must be psychologic as well as an attempt at eradicating or modifying the structural basis of pain.

101

The tests employed by practicing psychologists and by numerous pain clinics have their proponents and are too numerous to list and to evaluate here.

Patient pain drawing is a valuable diagnostic tool. By this technique the area of pain is clarified and is modified by descriptive shading or marking. Such a test facilitates communication otherwise obstructed by language barrier, educational differences, and discrepancy of medical terminology. In patients who are prone to magnify or even falsify their symptoms for whatever gains they see, the organicity or reasonableness of the symptoms is documented.

Pentothal interviews are also valued but used in a different concept of interview. In the somnolent state, as the patient is lightened from the induced hypnotic state (by way of intravenous pentothal, movements that induced pain in the wakeful state are performed and the patient's reaction noted. Authenticity of the symptom is given if there is the same response as invoked in the wakeful state, but emotional exaggeration is considered if there is a significant difference. All the precautions in using anesthesia are required so this cannot be performed as an office procedure.

The Minnesota Multiphasic Personality Inventory (M.M.P.I.) is a well-established profile evaluation of a person's personality. Its abuse as well as improper use must be avoided, but complete reliance or infallible interpretation cannot be assumed. The M.M.P.I. categorizes the patient at the time of the test but does not necessarily characterize that patient at the time of initial occurrence of illness or injury.

The Social Readjustment Rating Scale is also a valuable adjunct to evaluation of a patient. The time of disease or injury occurrence has been shown to relate to personal, social, vocational, and economic changes in the patient's life. High stress versus low stress are thus evaluated and may cast an insight into causation of the disability.

Personal interview by a competent and trained psychologist is highly desirable early in the intervention of the illness. When properly invoked, not only accurate diagnosis but also therapeutic intervention results.

The patient with chronic pain is entitled to have thorough medical evaluation to fully evaluate all the organic aspects of the symptoms, and the patient complaining of symptoms of pain must also be evaluated. Only in all of these aspects can total care of the patient with low back pathology be fully evaluated and treated.

SUMMARY

The diagram on the facing page summarizes patient care as outlined in this chapter.

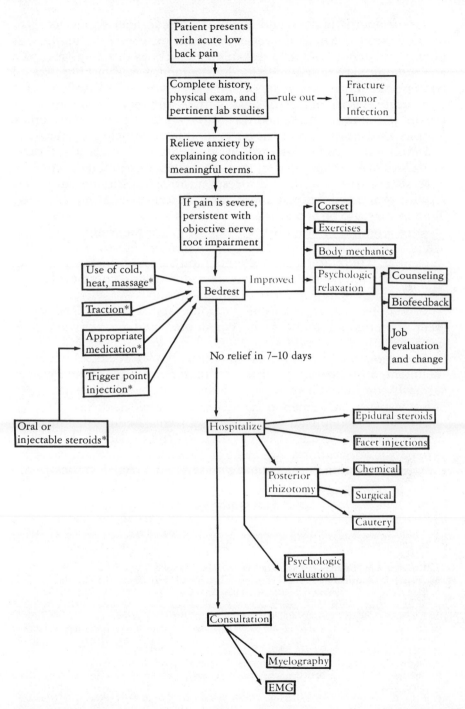

*May be used in hospital

A meaningful history and examination can lead to an accurate diagnosis when the functional anatomy of the vertebral column, the sites of pain production, and the mechanism and actions that may cause pain and disability are fully understood. Treatment based on these sound concepts should result in relief of pain and alleviation of disability.

An acute episode of low back pain with or without radicular leg pain is usually self-limiting. Time and rest of the affected part usually suffice. Various modalities such as trigger point injections, icing, medications, and even enforced bedrest may hasten recovery. Frequently various modalities may permit recovery without requiring cessation of activities.

Persistent pain that does not respond to simplistic measures may require prolonged bedrest specifically prescribed and implemented. Traction may also be beneficial.

After a period of seven to ten days of bedrest and especially with symptoms and signs of radicular irritation, more drastic measures are warranted. These are specifically epidural steroid injections of facet joints. After several such injections, provided that no contraindications exist, intramuscular or oral steroid administration is warranted.

Conditions such as piriformis syndrome, spinal stenosis, and severe facet arthritis must always be considered and proper treatment instituted.

After a reasonable period of time, usually two weeks, of adequate treatment, a myelogram preparatory for surgical exploration must be seriously considered if severe symptoms persist or there is evidence of objective nerve impairment or a progressive neurologic defect.

Chronic persistent pain with minimal objective findings and negative tests should always alert the physician to the possibility of serious psychologic implications. Adequate studies must be initiated early for complete evaluation of the patient with severe disabling low back pain.

REFERENCES

1. Falconer, M., McGeorge, M., and Begg, A.: Observations of the cause and mechanism of symptom production in sciatica and low back pain. J. Neurol. Neurosurg. Psychiat. 11:13–15, 1949.
2. Kellgren, J. H.: Deep pain sensibility. Lancet 1:943–949, 1949.
3. Forst, J. J.: Contribution a l'etude clinique de la sciatique. Paris These, No. 33, 1881.
4. Lasegue, Ch.: Considerations sur la sciatique. Arch Gen. Med 2 (Serie 6, Tome 4):558–580, 1864.
5. Lindahl, O.: Hyperalgesia of lumbar nerve roots in sciatica. Acta Orthop. Scand. 37:367, 1966.
6. Granit, R., Leksell, L., and Skoglund, C.R.: Fiber interaction in injured or compressed region of nerve. Brain 67:125, 1944.
7. McNab, I.: Chemonucleolysis. Clin. Neurosurg. 20:183–192, 1973.
8. Bertrand, G.: The "battered" root problem. Orthop. Clin. North Am. 6:305–310, 1975.
9. Shealy, C. N.: Facets in back and sciatic pain: A new approach to a major pain syndrome. Minn. Med. 57:199–203, 1974.
10. Rees, W. E. S.: Multiple bilateral subcutaneous rhizolysis of segmental nerves on the treatment of the intervertebral disc syndrome. Ann. Genet. 26:126–127, 1971.

11. Friberg, S., and Hult, L.: Comparative study of abrodel myelographic findings in low back pain and sciatica. Acta Orthop. Scand. 20:303, 1951.
12. Maigne, R.: Medical Orthopedics. Charles C Thomas, Springfield, IL, 1975.
13. Jonck, L. M.: The influence of weight bearing on the lumbar spine: A radiological study. So. African J. Radiol. Vol 2, No. 2, Deel 2, 25–29, 1964.
14. Hirsch, D., Ingelmark, B., and Miller, M.: The anatomical basis for low back pain. Acta Orthop. Scand. 33:1–17, 1963.
15. Berg, A.: Clinical and myelographic studies of conservatively treated cases of lumbar intervertebral disc protrusion. Acta Chir. Scand. 104:124–129, 1953.
16. Green, L. N.: Dexamethasone in the management of symptoms due to herniated lumbar disc. J. Neurol. Neurosurg. Psychiatry 38:1211–1217, 1975.

BIBLIOGRAPHY

Badgley, C. E.: The articular facet in relation to low back pain and sciatic radiation. J. Bone Joint Surg. 25:481, 1941.

Barton, C., and Neda, G.: Gravity lumbar reduction therapy program. Rehab. Pub. No. 751, The Sister Kenny Inst., Minneapolis, 1976.

Charnley, J.: Physical changes in the prolapsed disc. Lancet 2:43–44, 1958.

Cyriax, J.: Textbook of Orthopaedic Medicine. Cassell, London, 1955.

Edgar, M. A., and Park, W. M.: Induced pain patterns on passive straight-leg raising in lower lumbar disc protrusion. J. Bone Joint Surg. 563:658–667, 1974.

Epstein, J. A., et al,: Lumbar nerve root compression at the intervertebral foramina caused by arthritis of the posterior facets. J. Neurosurg. 39:362–369, 1973.

Epstein, J. A., et al.: Sciatica caused by nerve root entrapment in the lateral recess: The superior facet syndrome. J. Neurosurg. 36:584–589, 1972.

Falconer, M. A., McGeorge, M., and Begg, A. C.: Observations on the cause and mechanism of symptom-production in sciatica and low-back pain. J. Neurol. Neurosurg. Psychiatry 11:13–26, 1948.

Foerster, O.: The dermatomes in man, Brain 56:1–39, 1933.

Freiberg, A. H.: Sciatic pain and its relief by operations on muscle and fascia. Arch. Surg. 34:337–350, 1937.

Friberg, S. and Hult, L.: Comparative study of abrodil myelogram and operative findings in low back pain and sciatica. Acta. Orthop. Scand. 20:303–314, 1951.

Galante, J. O.: Tensile properties of the human lumbar annulus fibrosus. Acta Orthop. Scand. Supplement 100, 1967.

Gasser, H. S., and Erlanger, J.: Role of fiber size in establishment of nerve block by pressure or cocaine. Am. J. Physiol. 88:581, 1929.

Ghormley, R. K.: Low back pain with special reference to the articular facets with presentation of an operative procedure. J.A.M.A. 101:1773–1776, 1933.

Hallen, L. G.: The collagen and ground substance of human intervertebral disc at various ages. Acta Chem. Scand. 16:705–710, 1962.

Harris, R. I., and MacNab, I.: Structural changes in the lumbar intervertebral discs. J. Bone Joint Surg. 36[Br]:304–322, 1954.

Hartman, J. T., et al.: Intradural and extradural corticosteroids for sciatic pain. Orthop. Review 3:21–24, 1974.

Hendry, N. G. C.: The hydration of the nucleus pulposus and its relation to intervertebral disc derangement. J. Bone Joint Surg. 40[Br]:132–144, 1958.

Hockaday, J. M., and Whitty, C. W. M.: Patterns of referred pain in normal subject. Brain 90:481–496, 1967.

Inman, V. T., et al.: Referred pain from experimental irritative lesions. In Studies Relating to Pain in the Amputee. Series II, Issue 23, pp 49–78, June 1952.

Inman, V. T., and Saunders, J. B. deC. M.: Referred pain from skeletal structures. J. Nerv. Ment. Dis. 99:660-667, 1944.

Inman, V. T., and Saunders, J. B. deC. M.: The clinico-anatomical aspects of the lumbosacral region. Radiology 38:669–687, 1942.

Kelly, M.: Is pain due to pressure on nerves? Spinal tissues and the intervertebral disc. Neurology 6:32, 1956.

Krempen, J. F., Smith, B. S., and DeFreest, L. J: Selective nerve root infiltration for the evaluation of sciatica. Orthop. Clin. North Am. 6:311–315, 1975.

Landahl, C., and Rexed, B.: Histological changes in spinal nerve roots of operated cases of sciatica. Acta Orthop. Scand. 20:215–225, 1951.

Lindblom, K.: Technique and results of diagnostic disc puncture and injection (discography) in lumbar region. Acta Orthop. Scand. 20:315–326, 1951.

Maroudas, A., et al.: Factors involved in the nutrition of the human lumbar intervertebral disc: Cellularity and diffusion of glucose in vitro. J. Anat. 120:113–130, 1975.

Marshall, L. L. and Trethewie, E. R.: Chemical irritation of nerve root in disc prolapse. Lancet 2:320, 1973.

McCollum, D. E., and Stephen, C. R.: Use of graduated spinal anesthesia in the differential diagnoses of pain of the back and lower extremities. South. Med. J. 57:410, 1964.

Nachemson, A., et al.: In vitro diffusion of dye through the end plates and the annulus fibrosus of human lumbar intervertebral disc. Acta. Orthop. Scand. 41:589–607, 1970.

Pace, J. B., and Nagle, D.: Piriform syndrome. West J. Med. 124:435–439, 1976.

Pedersen, H. E., Blunch, C. F. J., and Gardner, E.: The anatomy of lumbosacral posterior rami and meningeal branches of the spinal nerves (sinu-vertebral nerves). J. Bone Joint Surg. 38[Am]:377–390, 1956.

Perey, O: Contrast medium examination of intervertebral discs of lower lumbar spine. Acta Orthop. Scand. 20:327–334, 1951.

Raaf, J.: Some observations regarding 905 patients operated upon for protruded lumbar intervertebral disc. Am J. Surg. 97:388–397,

Sarnoff, S. J., and Arrowwood, J. G.: Differential spinal block. Surgery 20:150, 1946.

Sehgal, A. D., et al.: Laboratory studies after intrathecal corticosteroids. Arch. Neurol. 9:64–68, 1963.

Sinclair, D. C., et al.: Intervertebral ligaments as source of segmental pain. J. Bone Joint Surg. 30[Br]:515–521, 1948.

Smyth, M. J., and Wright, V.: Sciatica and the intervertebral disc. An experimental study. J. Bone Joint Surg. 40[Am]:1401, 1958.

Sylven, B.: On biology of nucleus polposus. Acta Orthop. Scand. 20:275–279, 1951.

Troop, J. D. G.: Ph.D. Thesis, London University, 1968.

Winnie, A. P., and Collins, V. J.: Pain Clinic, I. Differential neural blockade in pain syndromes of questionable etiology. Med. Clin. North Am. 52:123–129, 1968.

Winnie, A. P., et al.: Pain Clinic, II. Intradural and extradural corticosteroids for sciatica. Anesth. Analg. 51:990–999, 1973.

CHAPTER 4

Neck and Upper Arm Pain

Pain in the neck, the head, or the interscapular area of the upper extremity (referred there from the neck) is second only to pain in the low back as a medical complaint.

Neck pain may be considered secondary to trauma or it may be appraised as unrelated to trauma. The type of trauma must also be defined. Conditions related to the neck, whether they are post-traumatic or unrelated to trauma, have many similarities but significant differences, and so their evaluation and their management must be clear.

FUNCTIONAL ANATOMY

As in all musculoskeletal diseases disabilities, and symptomatic entities, basic functional anatomy must be understood, examination must evaluate this function, and treatment must be related to functional recovery.

Functional anatomy has been fully discussed in my book, *Neck and Arm Pain*,[1] and, as it relates to the lower back, in Chapter 3, but a summary of pertinent functional anatomic and neurophysiologic aspects merits consideration here.

The cervical vertebrae, unlike their thoracic and lumbar counterparts, are more concave-convex and, in their motion, glide rather than rock. The facet orientation also differs in angulation (Fig. 77). All range of motion is possible in the cervical spine and there is specific localization of motion at various levels (Fig. 78).

At the occipito-atlas articulation only significant flexion and extension occurs, of approximately 10 degrees flexion (from neutral) and 25 degrees extension (Fig. 79).

Between the atlas and axis (C_1 and C_2), rotation occurs about the odontoid process with estimated rotation of 45 degrees to the left and 45 degrees to the right (Fig. 80). A slight degree of flexion and extension also

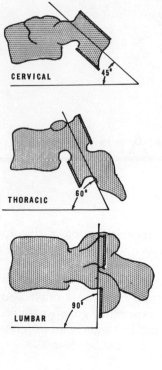

CERVICAL 45°

THORACIC 60°

LUMBAR 90°

FIGURE 77. Angulation of the cervical, thoracic, and lumbar facets.

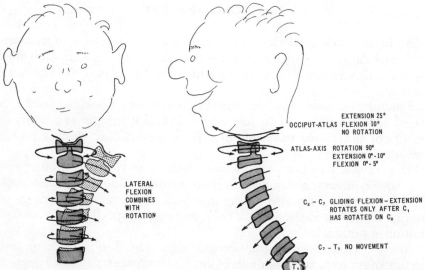

LATERAL
FLEXION
COMBINES
WITH
ROTATION

EXTENSION 25°
OCCIPUT-ATLAS FLEXION 10°
NO ROTATION

ATLAS-AXIS ROTATION 90°
EXTENSION 0°-10°
FLEXION 0°-5°

$C_2 - C_7$ GLIDING FLEXION–EXTENSION
ROTATES ONLY AFTER C_1
HAS ROTATED ON C_2

$C_7 - T_1$ NO MOVEMENT

T_1

FIGURE 78. Composite movements of the cervical spine with occipito-atlas flexion-extension but no rotation. There is 90 percent of rotation at the axoatlantoid joint and further movement of C_2 through C_7.

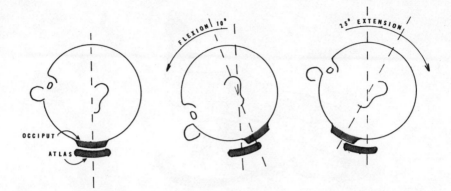

FIGURE 79. Motion possible at the occipito-atlas joint permitting 10 to 20 degrees flexion and 25 degrees extension, totaling 35 to 40 degrees of flexion-extension. No rotation possible.

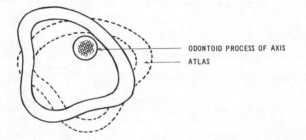

ODONTOID PROCESS OF AXIS

ATLAS

FIGURE 80. Motion of the atlas about the odontoid process of the axis, permitting rotation of 75 to 90 degrees.

occurs at this joint. Without motion elsewhere in the cervical spine, movement between the occiput and the axis (C₂) may permit 45 degrees of flexion-extension and 90 degrees of rotation.

Rotation of the second cervical vertebra upon the third cervical is limited by a bony locking mechanism in which the anterior tip of the upper articular process of the third cervical vertebra impinges upon the lateral process of the second cervical vertebra (Fig. 81). The lateral process of the vertebra is the lateral margin of the vertebral artery foramina. This locking mechanism prevents excessive rotation and thus protects the vertebral artery and the nerve root which descend the spinal nerve groove.

No intervertebral disks exist between the occipito-atlas and the atlantoaxial articulations and posteriorly the intervertebral foramina are *not* delineated by structure similar to the lower segments (Fig. 82). Stability of the atlantoaxial joint is afforded by ligamentous support, and instability is implied when in disease such as rheumatoid arthritis, which may weaken the ligamentous support.

109

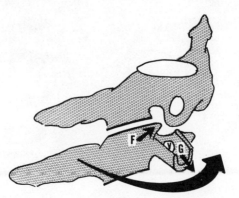

FIGURE 81.   Locking mechanism of C₂ upon C₃. Rotation of C₂ upon C₃ is limited by the mechanical locking of the articular structures. The anterior tip of the upper articular process of C₃ *(F)* impinges upon the lateral margin of the foramen of the vertebral artery *(V)*. G is the gutter through which emerges the nerve root C₃.

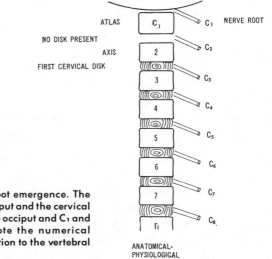

FIGURE 82.   Disk level of nerve root emergence. The nerve roots emerge between the occiput and the cervical spine, showing no disk between the occiput and C₁ and the atlas (C₁) and axis (C₂). Note the numerical derivation of the nerve roots in relation to the vertebral bodies.

Motion between subsequent cervical vertebrae (C₂ through C₇) has specificity that must be understood to evaluate normal motion and appreciate deviation from normal motion or injury that causes pain and disability.

Unlike other vertebral segments, i.e., thoracic and lumbar, the cervical vertebra has osseous elevations, of the posterolateral aspect that form pseudojoints termed uncovertebral joints (Fig. 83). The joints protect the contents of the spinal canal from intervertebral disk protrusion. They

110

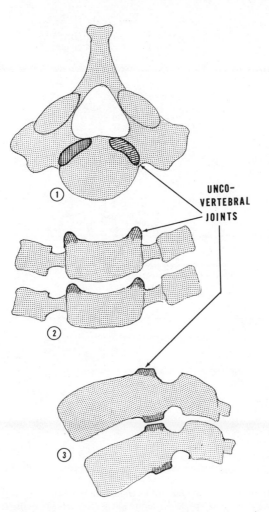

FIGURE 83. Uncovertebral joints. *1*, Superior view depicting the posterolateral placement of the uncovertebral joints; *2*, Anterorposterior view; *3*, Lateral view.

pose a pathologic problem in that they undergo hypertrophy and further calcification with disk degeneration and can encroach into the intervertebral foramen, causing root compression, or into the spinal canal, causing cord compression.

The functional unit includes the anterior weight-bearing portion and the posterior guiding-gliding portion (Fig. 84). The intervertebral disk is significantly broader anteriorly than posteriorly to accentuate the cervical lordosis. The nucleus is further anterior than in the lumbar spine with more posterior annular fibers, which protect the spinal canal contents from nuclear herniation (Figs. 85 and 86).

111

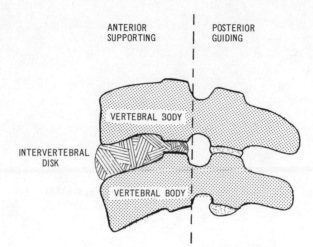

FIGURE 84. The functional unit in cross section.

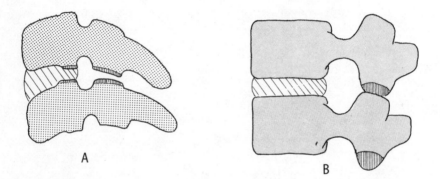

A

B

FIGURE 85. Comparative lateral views of cervical and lumbar functional units. A, Curved vertebral bodies of the cervical spine, the joints of Luschka (shown in the stippling), and the intervertebral foramina. B, Lumbar vertebra with different vertebral body contours and no joints of Luschka.

The musculature of the cervical spine is adequately clarified in textbooks of anatomy and kinesiology. I want to note here, however, that in early development the extensor muscles are the first to develop and remain the predominant muscles throughout life. In neonatal life the cervical extensor muscles develop early and are instrumental in forming the cervical lordosis. Until the individual assumes erect posture, these extensor muscles support the head against the forces of gravity. When the erect posture is assumed, the erector spinae muscles maintain the head erect and permit neck forward flexion by elongation or eccentric contraction. The flexors of the neck function primarily when force is

112

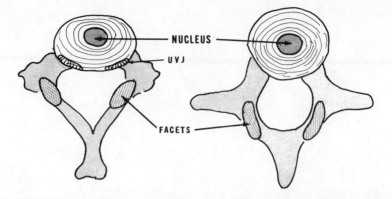

FIGURE 86. Comparison of cervical and lumbar vertebra. *Left*, Cervical vertebra with uncovertebral joints (UVJ), anterior-placed nucleus, and broader posterior annular fibers; *Right*, Lumbar vertebra with a centrally-placed nucleus and no uncovertebral joints. The angulation of the facets also differ as indicated.

applied to extend the neck or when gravity must be overcome in the supine position.

Motion is that of translatory gliding (Fig. 87). The disk deforms proportionately to allow this motion. Since the cervical facets are at a 45 degree angle with the frontal plane, any forward motion (flexion) causes the facets to glide on each other, thus elevating the posterior elements until ultimately the interspinous ligaments are stretched to their maximum and further flexion is stopped (Fig. 88).

The nuchal ligament is an elastic ligament comprised largely of collagen with viable amounts of elastin. The tissue is essentially avascular but is abundantly supplied by nerves which rise from the posterior primary division of the second, third, and fourth cervical vertebrae.

Recent experiments have postulated that some of the terminal nerve filaments act as proprioceptive receptors and are significant in attitudinal and tonic neck reflexes. This theory adds credence to the many bizarre neurologic symptoms of severe neck injuries.

Excessive force imposed upon the unit in flexion causes disruption of the interspinous ligaments and subluxation of the facet.

Extension of the cervical unit uses posterior gliding of the superior vertebra of the functional unit until the anterior angle of the inferior facet impacts upon the superior vertebra, forming a locking mechanism. In this normal action the intervertebral foramen is encroached upon but remains adequate for passage of the nerve root. However, excessive force in extension can result in fracture and instability, and excessive motion in either flexion or extension can violate the limitation and contract the foramina, spinal canal, and/or the vertebral arteries (Fig. 89).

Lateral motion and rotation, both occurring simultaneously, also

113

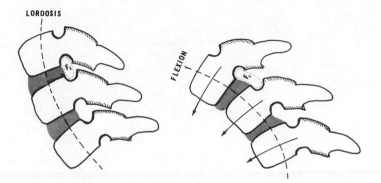

FIGURE 87. Translatory gliding. Forward flexion reveals the gliding of the superior upon the inferior vertebra with anterior disk compression, posterior disk separation, and opening of the intervertebral foramina upon flexion. Simultaneous closure occurs on extension.

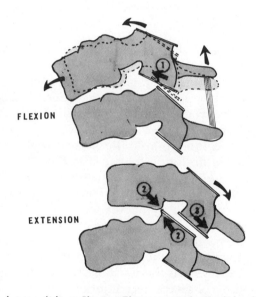

FIGURE 88. Translatory gliding. *Flexion:* The superior facet glides forward *(1)* and the essentially vertical facet elevates the posterior element until the interspinous ligament stops the movement. *Extension:* The superior facet glides posteriorly *(3)* until the inferior facet impinges upon the vertebra *(2)* and locks further extension movement.

affects the degree of foraminal opening. Lateral flexion to the right and rotation to the right causes closure of the right-sided foramina with opening of the left-sided foramina (Fig. 90). The opposite occurs in left lateral flexion and rotation to the left.

It becomes apparent that the greatest degree of facet approximation

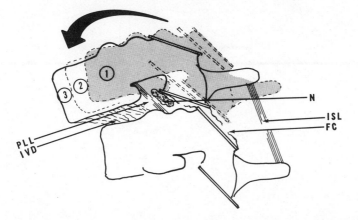

FIGURE 89.   Hyperextension-Hyperflexion Injury. Normal physiologic flexion (1 to 2) is possible with no soft tissue damage. When motion is exceeded (3), the intervertebral disk (IVD) is pathologically deformed and the posterior longitudinal ligament (PLL) strained or torn; the nerve (N) is acutely entrapped; the facet capsule (FC) is torn or stretched; and the interspinous ligament (ISL) is damaged.

and foraminal closure occurs in lateral motion with simultaneous extension of the neck. This fact has significant clinical application.

The nerves that emerge via the foramina are physiologically prevented from injury during physiologic motion by virtue of well-adjusted biomechanic features in the cervical spine. In forward flexion the spinal canal elongates, causing simultaneous foraminal opening. The emerging nerves change their angulation because the cord also elongates within the canal. The plastic spinal cord in enclosed in a plicated dura that elongates as the neck flexes. The pedicles bordering the foramina move away from the nerve roots as the foramina open.

In extension the canal length decreases and the foramina close. Entrapment of the emerging nerve roots is avoided by the shortening of the spinal cord and its dura, the dura plicating (Fig. 91), and the nerve root angulation approximating a right angle (Fig. 92).

In physiologic motion of normal spinal structure, no nerve root entrapment or injury occurs.

The emerging nerve roots into the foraminal canals are accompanied by dural sheath (Figs. 93 and 94), which contain spinal fluid, fat, and many unmyelinated somatic and sympathetic nerves. Encroachment or injury to these dural tissues can result in pain and dysfunction.

As the nerve roots leave the intervertebral foramen they branch into an anterior primary ramus and a posterior primary ramus (Fig. 95).

Since pain may occur from misuse, abuse, or injury to the cervical spine, the tissue sites from which pain may be initiated must be clarified.

The anterolongitudinal ligament is a relatively insensitive tissue. The

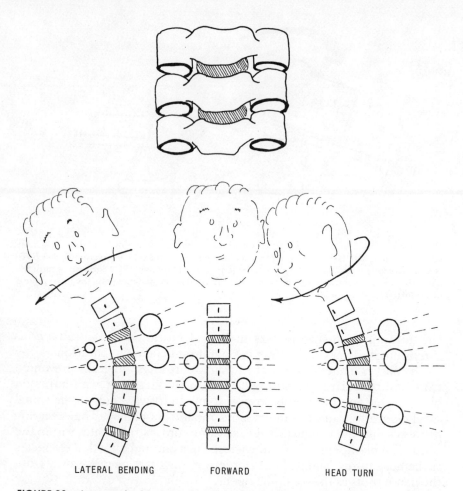

FIGURE 90. Intervertebral foraminal opening and closing on spine motion. Lateral bending closes the intervertebral foramina on the concave side and opens it on the convex side. Rotation, head turning, causes closure on the side towards which rotation occurs and opening on the side away from which rotation occurs.

LATERAL BENDING   FORWARD   HEAD TURN

vertebral bodies can evoke pain in conditions such as metastatic invasion, fracture, or osteomyelitis, but here the complaint is that of a deep, nonspecific, continuous pain, usually not related to motion.

The intervertebral disk in the adult is avascular and without nerve penetration of its substance so it must be considered insensitive. This remains controversial, however, since Cloward[2] claimed that, in performing cervical diskography, no pain resulted if the disk nucleus was intact and the amount injected was small (0.2 to 0.3 cc.). Pain could result if there were any nuclear tears. Holt[3] disagreed, claiming pain resulted in *any* disk injection. It was also claimed that pressure on the superficial part

116

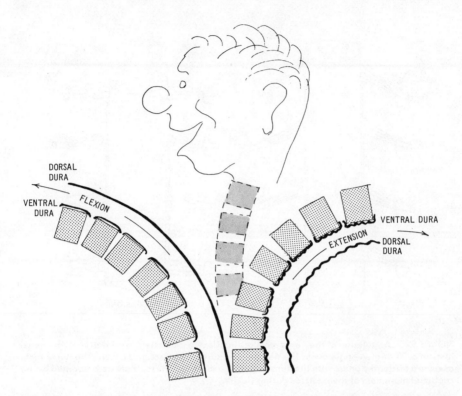

FIGURE 91. Length of spinal canal and dural sheath. Flexion elongates the spinal canal causing full elongation of the dural sheath. Extension shortens the length of the spinal canal causing plication of the dural sheath.

of the disk in the posterior or posterolateral regions caused lateralized pain with projection into the shoulder and upper arm region. Pressure upon the anterolateral portion of the disk caused radiation into the interscapular musculature. This interscapular pain radiation is identical to the referred pain from irritating the motor root. The exact sensitivity of the intervertebral disk remains unclear.

The posterolongitudinal ligament is exquisitely sensitive, being innervated by the recurrent nerve of Luschka. The nerve root within the foramen, also innervated by the recurrent meningeal nerve, is very sensitive. The posterior joints, the facets, are synovial joints and are highly innervated and sensitive. The posterior neck muscles, like all skeletal muscles, can be the site of pain and tenderness. It is evident that pain must originate in the area of the foramina and the posterior articular elements of the functional unit, thus it behooves the clinician to determine the injury, illness, motion, or position of the neck that irritates the susceptible tissues. The history and physical examination demands verification of mechanism and tissue site of irritation.

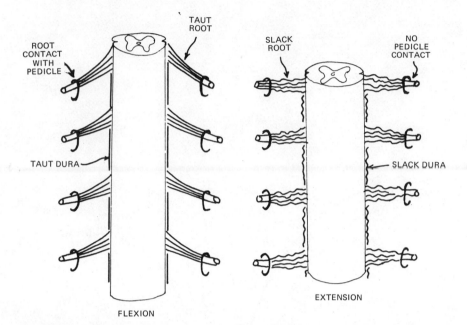

FIGURE 92.   Angulation of the nerve roots during flexion and extension. *Extension:* The nerve roots at a 90 degree angle avoid the descending part of the pedicle. *Flexion:* The nerve roots assume a 60 degree angle with the dura becoming more taut. The roots are accompanied by the cephalad movement of the pedicles during flexion.

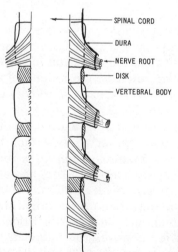

FIGURE 93.   Nerve roots with dural sheath. The fila which form the nerve root emerge from the spinal cord and are accompanied through the interverte-bral foramina by the dural sheath.

118

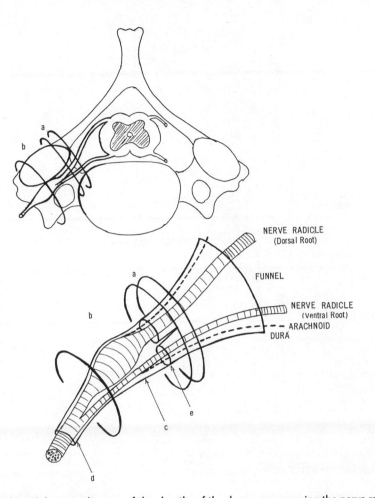

FIGURE 94.   Schematic diagram of the sheaths of the dura accompanying the nerve roots in
the intervertebral foramina.
A—intervertebral foramen
B—gutter of the transverse process
C—attachment of the dura
D—nerve with only dural sheath
E—apex of funnel

Pain in the neck or referred from the neck must be related to irritation
of these sensitive tissues. The manner in which this pain and resultant
disability occurs is elicited by history and examination. As the site of
tissue injury and the mechanism of irritation becomes apparent, a
diagnosis evolves and meaningful treatment results.

Injury to the neck may be classified as acute, recurrent, or chronic.
Trauma may be further divided into external force application, postural
trauma, or trauma as a resultant of anxiety, tension, or emotions. Details

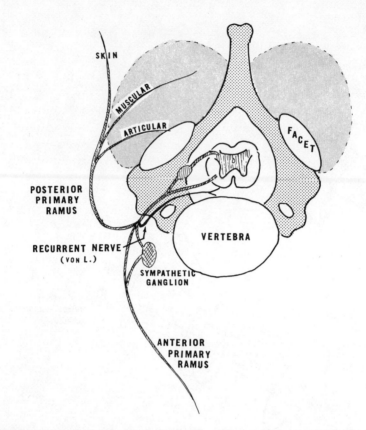

FIGURE 95. Innervation of the cervical roots: The posterior primary ramus divides into skin, muscular, and articular branches; the anterior primary ramus, with a sympathetic innervation, proceeds to the dermatome areas.

of the injury are related by the history; the examination confirms the injury; and both are sustained with the clinical evaluation which clarifies the soft tissue injury and possible neurologic involvement.

ACUTE TRAUMA

External forces imposed upon the neck can vary from an automobile accident, an athletic injury, a fall, to a direct blow. The extent of injury cannot always be immediately determined and the primary care of the injured patient must be scrupulously observed.

At the scene of an accident it must be assumed that the neck has been injured and possibly the cord has also been traumatized. This is especially true if the patient is unconscious, has sustained a simultaneous forehead laceration, or complains of neck pain. In a severe or potentially severe

120

injury, the patient must be moved with the utmost care. The head and neck must be immobilized manually to avoid any flexion or extension. A slight degree of traction is permissible but should be merely to assure straight alignment of the neck and immobilization.

Roentgenograms, which are mandatory in severe injuries, should be taken early, and, while being taken, the patient's head and neck should be immobilized by the physician. The decision as to which specialist (orthopedist or neurosurgeon) is best for the care of the patient requires choosing the physician who has the greatest experience in neck injuries.

An adequate airway must be maintained while the patient is still at the place of injury. The usual oral airway requires hyperextension of the patient's neck for insertion which cannot be permitted in possible fracture-dislocation of the neck with possible cord embarassment. A nasal pharyngeal latex airway is lifesaving and can be inserted simply and safely. No analgesics should be given the injured patient because they depress alertness and the interpretation of pain and sensation.

Further care of the cervical fracture-dislocation with or without neurologic deficit is beyond the scope of this text and requires the intercedence of a specialist. The precautions listed here are to insure the patient's reaching this specialist without additional damage or injury.

In injuries with no severe impact or symptoms and no neurologic deficit the condition of acute sprain with soft tissue injury must be assumed to have occurred. An accident need not be severe to cause cervical injury; in fact, injury can be sustained from rapid braking of a car, stepping from an unnoticed curb, or stepping into a depression in the ground.

Motion of the various segments of the spine are specific and limited. Physiologic range of motion can be exceeded by external force. The atlas on the axis (C_1 on C_2) is precariously balanced and does not permit much flexion-extension. External force can exceed this normal range and cause vascular injury as well as acute spinal cord trauma.

Subsequent functional units below the level of the second to eighth cervical vertebrae (C_2 to C_8) flex and extend to the extent of articular (facet) subluxation, especially when the head is rotated. Excessive force or a force applied when the patient is least alerted to this imposed force can overstretch the ligamentous capsular tissues that normally prevent subluxation.

Forceful flexion and/or extension causes damage to the longitudinal ligaments, the intervertebral disk, the articular capsules, the ligaments, and the muscles of the neck. Simultaneously the spinal canal is acutely narrowed as are the intervertebral foramina (Fig. 96). The sensitive tissues of the functional unit are involved, which results in local and referred pain.

121

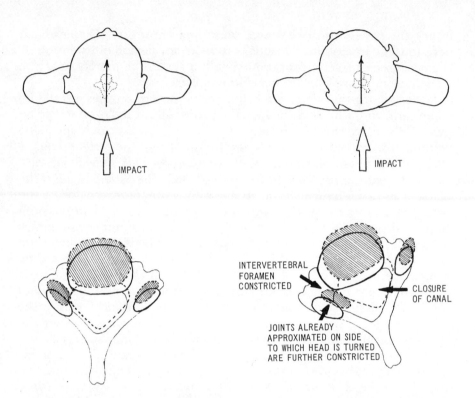

FIGURE 96. Hyperflexion injury with head facing forward versus effect of head rotation. An impact from the rear when head is turned causes further intervertebral foramen closure, excessive spinal canal closure, and excessive joint subluxation. This explains severity of injuries when a car occupant is struck when his/her head is turned to left or right.

Diagnosis and Evaluation

When roentgenograms fail to reveal a fracture or dislocation and neurologic deficit is not present, the extent and specification of the involved soft tissues need to be determined. This requires a complete neurologic examination of the peripheral nerves and of upper motor symptoms related to cord injury and a careful examination of soft tissue.

Active and passive range of motion must be evaluated. Normally active flexion allows the patient to place his chin on his sternum, extension permits the patient to look up vertically, rotation allows the chin to touch the acromioclavicular joint, and lateral flexion permits the ear to touch the shoulder. These motions are best tested with the patient seated in a chair or on the examining table. Passive examination is best done with the patient in the supine position and relaxed. Flexion must differentiate flexion of the head upon the cervical spine and flexion of the total cervical spine (Fig. 97).

122

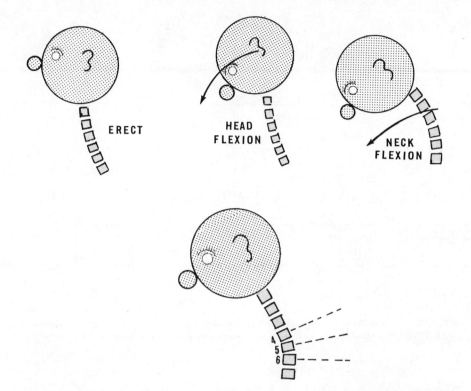

FIGURE 97. Head and cervical spine flexion. In head flexion the head is flexed upon the cervical spine with movement only at the occipito-atlas; in neck flexion there is reversal of the cervical lordosis. Most flexion occurs between C₄ and C₅ or C₅ and C₆.

Lateral flexion and rotation occur simultaneously when done actively by the patient.

Recording of this examination by exact measurement in numerical degrees is difficult to standardize. The method of Maigne is practical and meaningful (Fig. 98). An X may be used when motion is limited but not painful.

Tenderness must be carefully evaluated. Direct pressure on the posterior superior spine will localize specific site of involvement. This pressure can be direct or lateral (Fig. 99).

Pressure over the interspinous ligament can be determined also. Paraspinous tenderness or finding muscular nodules again is of value for localization. I have not found feeling a spasm in muscles to be accurate or revealing.

Since a hyperextension injury frequently overwhelms and, thus, injures the flexors of the neck (Fig. 100), various symptoms are frequent. Often the patient complains of difficulty swallowing, of pain and

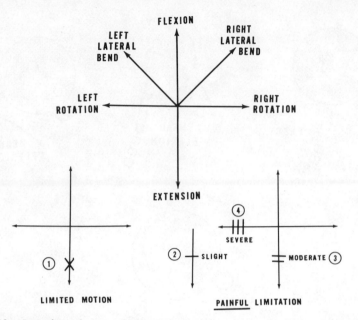

FIGURE 98. Recording of range of motion of cervical spine. Lines are placed for restriction and simultaneous pain and X is utilized with mere limitation and no pain.

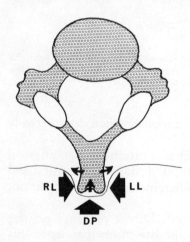

FIGURE 99. Methods of movement of the vertebral segment with direct pressure *(DP)* or right lateral *(RL)* or left lateral *(LL)* pressure upon the posterior superior spine.

discomfort in the neck flexor muscles, and of difficulty in lifting the head from the bed or pillow from a supine reclining position.

Treatment of Acute Injury

Because an acute injury may cause soft tissue inflammation with undoubted microscopic tissue damage, these tissues must be placed at

124

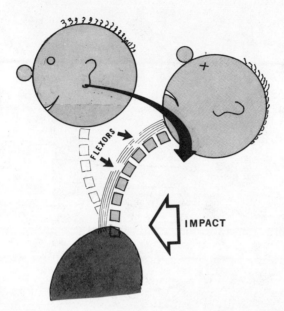

FIGURE 100. Neck flexor trauma from rear-end injury. Hyperextension causes an overstretch and inappropriate contraction of the neck flexors with residual flexor disability.

rest in a physiologic position. A muscle spasm follows an acute injury to protect the injured part by immobilization, but this spasm, as well as being beneficial, may also be detrimental. Therefore, proper immobilization of the neck is mandatory but presents problems.

Immobilization of the neck requires prevention of motion of the occipito-cervical junction and of the cervical column from the second to seventh vertebra. Rotation and lateral flexion must also be restricted. No specific splinting device has as yet been devised to perform all of these criteria.

COLLAR AND BRACE. Cervical collars or braces have been shown to be relatively ineffectual in immobilizing the head and neck. The felt collar depicted in Figure 101 is made of felt ¼ inch to ½ inch thick contoured to the individual's neck. The enclosing stockinette also encircles the chin and partially immobilizes the occipital cervical juncture. This stockinet-felt collar supports the head, is comfortable, and encourages a slight degree of neck flexion. At best it restricts merely 5 percent to 10 percent of flexion-extension, lateral bending, and rotation. Its main advantage is head support, comfort, and minimal expense.

Plastic collars are made more firm than soft felt collars by including metal strips that adjust the height of the collar. It is estimated this collar restricts flexion and extension by 75 percent but restricts rotation by only 50 percent.

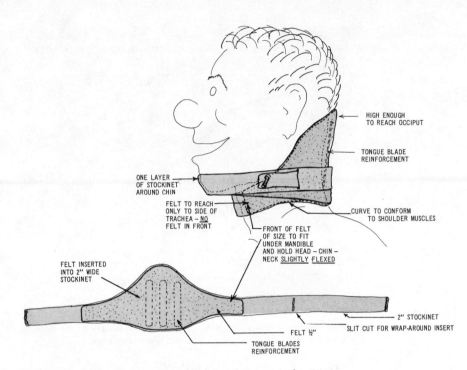

FIGURE 101. Felt collar recommended for occipito-cervical restriction.

The four-poster cervical brace is cumbersome to maintain and difficult to adjust. It restricts approximately the same degree of flexion-extension as does the plastic collar but permits rotation of the atlantoaxial joint.

More recently a brace called the Guildford brace (Fig. 102) has been shown to restrict flexion extension *and* rotation by 90 percent. It is comfortable and easy to wear. It must be individually fitted to the patient and immobilizes *all* the occipito-cervical joint movements. It has all the immobilizing virtues of a Minerva plaster jacket without the cumbersome and cosmetic deterrents. As the need for the total immobilization decreases it can be removed for other forms of treatment.

No collar or brace has proven to fully immobilize the neck in all directions. Only a Halo body jacket brace will fully immobilize the neck in flexion-extension, rotation, and lateral bending, and this is regarded only in severe orthopedic or neurologic problems.

EXERCISES. Exercises should be considered very early in the care of hyperextension injuries of significant intensity of the cervical spine. The initial weakness of neck flexors does not permit the patient to elevate the head from the pillow; therefore the flexors do not function to stabilize the neck nor do they reciprocally relax the antagonistic extensors, which

126

are frequently in protective spasm. Neck flexor weakness persists after subsidence of the acute aspects of the injury and has been incriminated as a cause of persistence of neck pain and dysfunction. Many cases of neck pain not related to acute trauma have also revealed weakness of neck flexion.

Exercises must be both isometric and isotonic and must be for short flexors (occipito-cervical) and long flexors (cervical). Initially exercises have to be active/assistive with a therapist or a trained member of the family administering the exercises. At first, isometric exercises which require *no* joint motion are desirable and performed with varying degrees of rotation to the left and the right. Gradually, isotonic exercise with increasing resistance should be initiated to improve range of motion as

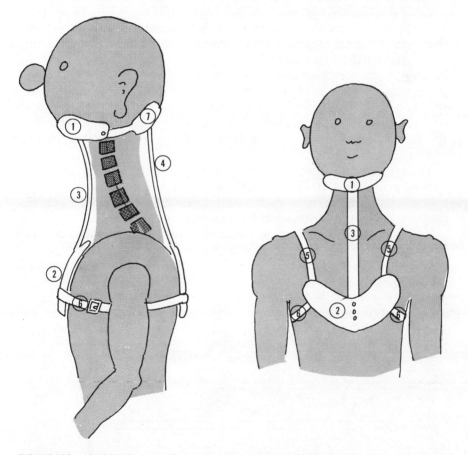

FIGURE 102. Guildford cervical brace. Brace immobilizes the head for flexion-extension and rotation by chin rest *(1)* and occipital pad *(7)* which can be narrowed or lengthened to fit. Chest pad *(2)* is secured to rib cage by shoulder straps *(5)* and chest straps *(6)*. Upright bars *(3 and 4)* adjust the flexion and extension of the spine to determine the degree of flexion.

well as strength. Posture training to realign the cervical physiologic lordosis should also be considered.

PHYSICAL THERAPY. Modalities such as heat, ice, ultrasound, and infrared have their advocates, but usually ice early and heat in later stages of recovery are preferred. Massage of the painful, tender, and endurated muscles is of value preceding gentle gradual exercises.

TRACTION. Traction also has equivocal support but has clinical value when properly applied. The force, direction, and duration of traction again is controversial. DeSeze and Levernieux[4] determined that 260 pounds of traction produced approximately 2 mm. of separation between the fifth and sixth cervical and the sixth and seventh cervical vertebrae and 400 pounds of traction produced 10 mm. separation between the fourth cervical and first thoracic vertebrae. None of these forces are realistic nor clinically applicable.

Traction with neck in slight flexion will decrease the lordosis and open the intervertebral foramina and separate the posterior articulations. Crue[5] reported an increase of 1.5 mm. in the vertical diameter of the intervertebral foramina of the fifth to sixth cervical vertebrae with 5 pounds of traction applied to the neck at 20 degrees of flexion for a period of 24 hours. Colachis[6] deducted from his studies that 30 pounds of traction applied for 7 seconds increased the separation according to the flexion angle. Twenty degrees in the author's opinion is the optimum angle for subjective relief (Fig. 103).

The reclining position is optimum but traction properly applied in the sitting position can also be effective. In severe acute problems, continuous reclining traction for two to three days is effective. In chronic cases, home traction used several times a day is effective.

Traction can also be applied manually which can be combined with lateral and rotatory stretch to increase range of motion.

Continuous versus intermittent traction remains unresolved insofar as

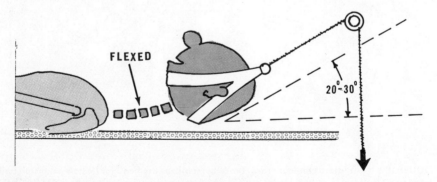

FLEXED

20°-30°

FIGURE 103. Cervical traction applied to the supine patient causing cervical spine flexion with the angle of pull between 20 and 30 degrees.

its relative effectiveness. Some patients feel able to relax better with intermittent traction. In some, the undulating traction initiates stretch reflexes and causes some muscular resistance to the traction. Both types of traction are of value and depend upon the patient acceptance and the claimed benefit to the symptomatology.

Cervical traction is readily applied in the clinic set-up, but for the patient this requires hours of driving, difficulty parking, and often long waits in the clinic for treatment. The trip home, in addition to all these inconveniences, frequently nullifies the benefit of the traction treatment. Once the traction has been tried upon the patient and the reaction evaluated, a home traction apparatus and regime is desirable.

The usual cervical traction equipment applied to the door is ineffective (Fig. 104). A more effective home traction is to suspend the traction rope from a chinning bar that can be fitted in any doorway and is removable (Fig. 105). The amount of traction can be varied by the patient and the angle of pull can be reasonably established.

After subsidence of acute symptomatology by the use of a collar,

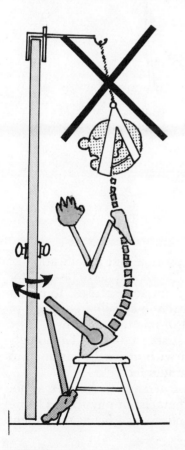

FIGURE 104. Ineffective home door cervical traction. The patient is too close to the door to get the correct neck flexion angle. The door freely opens and closes, not permitting constant traction. The patient cannot extend the legs or assume a comfortable position. This type of home traction is not recommended.

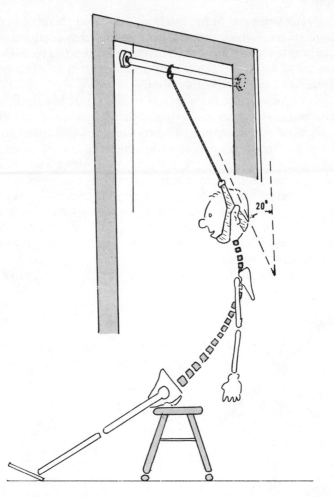

FIGURE 105.   Recommended home traction from chinning bar in the sitting position.

concomitant exercises and daily traction posture and modification of daily activities must be undertaken. The collar should never be used for prolonged periods nor be used without appropriate exercises. To persist with a collar leads to contracture of soft tissue and limited range of motion, muscular atrophy, and psychologic dependence. The exercises—at first isometric, and with assistance—should gradually evolve into active isotonic exercises and with gradual resistance to increase range of motion and strength and endurance.

Complete range of motion or even increase in range of motion may not be possible with skilled assistance from a therapist or physician. Gentle stretching and gentle manipulation may be necessary. Having the patient actively assist in these exercises (termed active/assistive exercise) accomplishes a more rapid gain of range and painfree function.

POSTURE. Posture, previously discussed in Chapter 3, requires attention to the entire body alignment. Lumbar lordosis decreased by pelvic tilting realigns the vertebral column to the center of gravity and thus centralizes the head above the cervical spine. The distraction exercise depicted in Figure 106 is an excellent method of improving posture and maintaining proper posture during daily activities.

Posture per se can be the trauma that initiates cervical or radicular pain. The forward head posture on the basis of a round back posture causes an increase in cervical lordosis with foraminal closure and increases approachment of the posterior articulation (facets).

Posture is based on several etiologic factors: (1) familial or evolving from habits of adolescence, (2) a manifestation of disease such as parkinsonism or rheumatoid spondylitis, or (3) a depiction of the emotions. Depression is the most prevalent emotional condition that

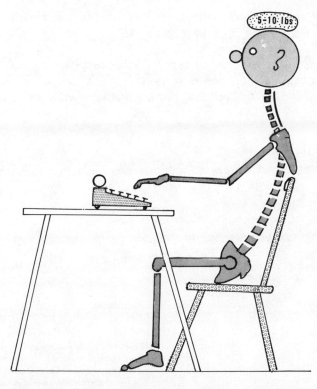

FIGURE 106. Distraction exercise for posture training. With a weight of 5 to 10 pounds within a sandbag upon the head, the posture is maintained erect and the cervical lordosis is minimal. Proprioceptive concept of posture is learned with no effort.

causes the round-shoulder posture. Depression also impairs muscle tone and muscular activity, which further impairs good erect posture.

The "round-back forward-head posture" causes the head to be anterior to the center of gravity. The weight of the head thus has the equivalent weight of 10 pounds plus the number of inches ahead of the center of gravity (Fig. 107). Usually this can be 10×3 or 10×4 or the equivalent of 30 to 40 pounds. To maintain direct vision, the head must be extended upon the cervical spine and thus increases the cervical lordosis. By virtue of the increased lordosis, the intervertebral foramina are compressed and the posterior articulations become more weight bearing. The erector spinae (cervical) muscles contract to support the cantilevered head and thus are in a constant state of isometric shortened contractions and contribute to pain and disability. Thus, the foraminal closures and the increase weight bearing of the facets cause local pain and radiating pain.

Medical and psychiatric treatment of depression can greatly influence the emotional status of the patient, but the postural component of depression is too frequently ignored and presents as a painful residual. Conversely, a forward head posture attributed to a round back is frequently treated as postural with complete unawareness of the underlying depression as an aggravating factor (Fig. 108). They must be considered and treated together.

Persistence of the increased cervical lordotic posture can become an aggravating cause of degeneration spondylosis. The uncovertebral joints are normally in direct opposition but are not significantly weight bearing

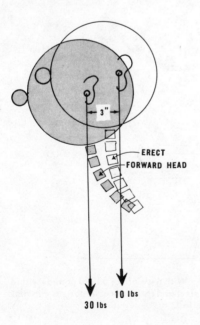

ERECT
FORWARD HEAD

10 lbs

30 lbs

FIGURE 107. With erect posture the weight of the head (approximately 10 pounds) is maintained directly above the center of gravity. In a forward head posture the head is approximately 3 inches in front of the center of gravity, causing the head to become an estimated 30 pounds of weight upon the cervical spine.

132

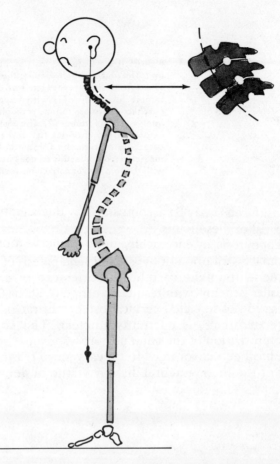

FIGURE 108. Forward head posture of depression. The depressed person has a rounded back posture with forward posture that increases the cervical lordosis. This increased lordosis approximates the posterior articulations and narrows the intervertebral foraminae (*right above*).

nor do they manifest friction on motion. With increased lordosis the uncovertebral joints are exposed to compression and irritation and thus undergo exostosis.

DEGENERATIVE JOINT DISEASE

As the hypertrophied uncovertebral joints develop, they protrude posteriorly to encroach into the intervertebral foramina and thus encroach upon the nerve roots and also the spinal canal to compress the cord (Fig. 109). These osteophytes may be accompanied by cervical disk herniation and, because of the approximation of the posterior

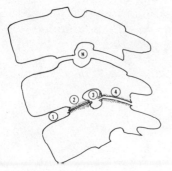

FIGURE 109. Osteophytosis of the cervical functional unit. The above unit *(N)* depicts the normal relationship. The unit below reveals a narrowed degenerated intervertebral disk *(1)* with hypertrophied joints of Luschka. There are formed osteophytes protruding into intervertebral foramen *(2)*. The posterior facets undergo osteoarthritic changes *(3)* and contribute posterior osteophytes into the foramen *(4)*. Local pain and restricted motion results as does nerve root compression with the foramen or cord compression within the canal.

articulations (facets), may be accompanied by degenerative joint disease (osteoarthritis) of these joints.

The osteophytosis, by encroaching upon the nerve root, can cause pain of dermatomal distribution or hypasthesia and paresis of that nerve root. Encroachment upon the vertebral arteries can result in increasing vertebrobasilar ischemia with resultant vertigo or ataxia (Fig. 110).

The causation of so-called cervical osteoarthritis, or more properly called osteoarthrosis, is currently unclear. The concept of slow progressive protrusion of the annulus fibrosus giving rise to osteophyte formation has been advocated. It appears more feasible that gradual narrowing of the intervertebral disk by virtue of dehydration causes

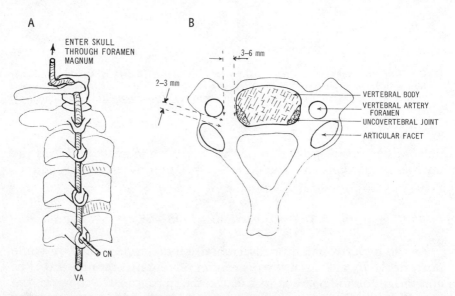

FIGURE 110. Passage of the vertebral artery through the cervical spine. *A*, Angulation at the C_1 to C_2 segment. *B*, Relationship to the uncovertebral joints and the facets.

134

approximation of the posterior articulations and the uncovertebral joints. As a resultant slackening of the longitudinal ligaments, the protruding disk material gradually forms the osteophyte (Fig. 111).

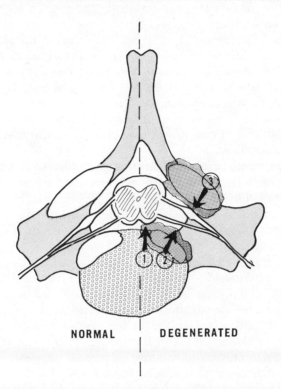

NORMAL | DEGENERATED

FIGURE 111. Osteoarthrosis of the cervical spine. Left side reveals normal facet joint and uncovertebral joint with no encroachment upon the nerve root or cord. Right side shows osteophytosis of uncovertebral joint encroaching into spinal canal (1) and into intervertebral foramen (2). Degenerative changes of facet joint (3) encroaches into foramen. All three osteophytes can cause neurologic deficit of cord or nerve root.

HEADACHE

Headache is defined as a discomfort in the forehead and scalp area excluding the face below the eyebrows. This area is supplied by the great occipital nerves which are derived from the second cervical roots and by the trigeminal nerve.

Headache may occur from excessive neural discharge originating from intracranial, cranial, or extracranial tissues. Discharges may result from abnormal psychologic reactions mediated via muscular contraction. Headaches occur frequently and may vary from annoyance to severe, prolonged, and recurrent incapacitation. It has been estimated that 6 to 7 percent of the population suffer from chronic headache that causes loss of time from work or school.

Intracranial headaches may be acute, serious, and even life threatening. Extracranial headaches are predominantly chronic, do not endanger life, and may be very unpleasant.

Cranial and intracranial headaches have numerous causations that require thorough neurologic or neurosurgic investigation and treatment. Headaches that persist and do not respond to conservative measures or that elicit unusual symptoms and/or findings should be referred for thorough neurologic consultation. The majority of chronic headaches encountered in sophisticated communities are not due to organic disease and are resolved by evaluation using clinical methods rather than by laboratory procedures.

Most extracranial headaches have been attributed to vascular origin. All large vessels (intracranial or extracranial) are sensitive to pain produced by distension or displacement. Extracranial headaches of the episodic migraine type are related to the vessels or the muscles of the scalp. Headache, however, especially in middle age or later, is the most common presenting symptom of cervical spondylosis.

Headache resulting from cervical spondylosis is usually occipital and unilateral. It is frequently claimed to start in the back of the neck, spread up to the occipital region, and often to the forehead or the eye. Pain is usually characterized as nagging and wearing rather than the throbbing or bursting so frequently claimed in headaches of vascular origin or headaches of increased intracranial pressure. There is often relationship to posture and aggravation by active or passive movement of the neck. The exact mechanism of headache from cervical spondylosis remains obscure, but relationship to the posterior articulation with the cervical muscles must be conceded because so many respond to conservative measures directed to these tissues.

Several concepts emerge that may elucidate the mechanism of headache from the cervical spine. The greater and lesser occipital nerves of the second and third cervical root derivation emerge from the cervical spine without the benefit of bony canal protection. There are no intervertebral disks between the occiput and the atlas (C_1) nor between the axis and the atlas (C_2). The bony configuration of these two cervical vertebrae also do not form the spinal nerve canal observed in subsequent vertebral functional units. The two nerve roots (C_2 and C_3) forming the occipital nerves emerge through ligamentous and muscular tissue towards their final destination to the occipital areas (Fig. 112).

All injuries to the cervical spine cause spasm, injury, or inflammation of the erector spinae muscles as well as to myofascial connection to the cranial periosteum. The nerve roots can well be entrapped within these irritated tissues. This would clarify the benefit derived from heat,

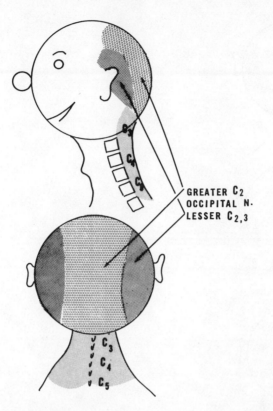

FIGURE 112. Dermatomal distribution of the occipital nerves. The greater occipital and lesser occipital nerves formed from roots C₂, C₃, and C₄ refer pain to the occipital vertex or parietal areas of the head as depicted by shaded areas.

massage, traction, manipulation, or injections to the incriminating tissues.

Headache referred from lower cervical areas, that is, third, fourth, and fifth, can be assumed to originate from the posterior articulation (facets) which irritate the posterior primary division whose ultimate distribution is to the occiput. Research has revealed that injecting irritating substance into the muscles adjacent to cervical levels has shown pain radiation into areas clinically consistent with headache patterns (Fig. 113).

Trigger areas with the sternocleidomastoid, splenius capitis, temporalis, masseter, or trapezius muscles can refer pain to occipital areas (Fig. 114). These occipital areas also respond therapeutically to injection of these trigger areas.

That headaches can occur from cervical osteophytes may be clarified only if an added trauma is superimposed upon the abnormal functional unit. The mere presence of cervical diskogenic disease with osteophytosis is usually asymptomatic and results in pain from trauma,

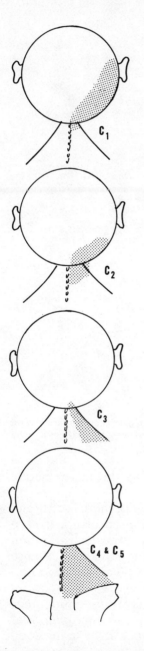

FIGURE 113. Referred zones of root levels. Injection of an irritant into paraspinous areas of the cervical spine (C_1 through C_5) results in pain noted by patients in the shaded areas.

postural stress, or emotional tension. Benefit so frequently obtained from traction and manipulation can be attributed to releasing the nerve root encroachment within the foramen or from decreasing the facet joint irritation to the posterior primary division of the nerve roots.

Long-sustained contraction of the skeletal muscle of the cervical spine,

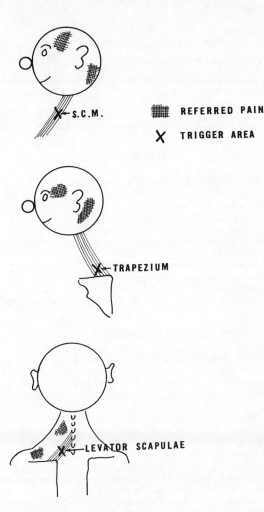

FIGURE 114. Trigger points and referred pain. Tender areas in various areas of the neck and shoulder when irritated can refer pain to distal sites.

the face, and the scalp can cause so-called tension headaches. These are also headaches described by the patient as vice-like and are noted in the temporal, occipital, parietal, or frontal areas.

Tension headaches fall into the symptom groups of psychosomatic disorders in which life stresses, unsolved inner conflicts, anger, and so forth lead to physical symptoms that in turn aggravate the anxiety.

Psychotherapy, behavior therapy, and assertiveness training help the patient deal more effectively with inner conflicts. Recently an adjunct to psychologic treatment and physical therapy modalities has been biofeedback.

Biofeedback employs electromyograph which records muscular contraction and permits the patient, once aware of the muscular tension, to learn to reduce muscular tension and thus reduce the related muscular pain. There are numerous methods of reducing muscular contraction (tension) such as progressive relaxation, meditation, self hypnosis, and yoga-type exercises about which there is voluminous literature.

REFERENCES

1. Cailliet, R.: Neck and Arm Pain, F. A. Davis Co., Philadelphia, 1964.
2. Cloward, R. B.: The clinical significance of the sinuvertebral nerve of the cervical spine in relation to the cervical disc syndrome. J. Neurol. Neurosurg. Psychiatry 23:321, 1960.
3. Holt, E. P.: Fallacy of cervical discography. J.A.M.A. 188:799, 1964.
4. De Seze, S., and Levernieux, J.: Les tractions vertebrales; priemeres etudes experimentales et resultats therapeutiques d'apres une experience de quatres annes. Semaines des Hopitaux (de Paris) 27:2085–2104, 1951.
5. Crue, B. J.: Importance of flexion in cervical traction for radiculitis. U.S.A.F. Med. J. 8:375–380, 1957.
6. Colachis, S. C., and Strohm, B. R.: Cervical traction: Relationship of traction time to varied tractive force with constant angle of pull. Arch. Phys. Med. Rehabil. 46:815–819, 1965.

BIBLIOGRAPHY

Birk, L. (ed.): Biofeedback: Behavior Medicine. Greene & Stratton, New York, 1973.
Blumenthal, L.: Injury to the cervical spine as a cause of headache. Postgrad. Med. 56:147–153, 1974.
Braaf, M. M. and Rosner, S.: Trauma of the cervical spine as cause of chronic headache. J. Trauma 15:441–446, 1975.
Brain, L.: Some unsolved problems of cervical spondylosis. Br. Med. J. 1:771–776, 1963.
Caldwell, J. W., and Krusen, E. M.: Effectiveness of cervical traction in treatment of neck problems: Evaluation of various methods. Arch. Phys. Med. Rehabil. 43:214–222, 1962.
Cloward, R. B.: The clinical significance of the sinu-vertebral nerve of the cervical spine in relation to the cervical disc syndrome. J. Neurol. Neurosurg. Psychiatry 23:321, 1960.
Colachis, S. C., Jr., and Strohm, B. R.: Effect of duration of intermittent cervical traction on vertebral separation. Arch. Phys. Med. Rehabil. 47:353, 1966.
Coppola, A. R.: Disease of the cervical spine and nerve root pain. Va. Med. Mon. 101:199–201, 1974. (A concise report on the symptoms, diagnosis, and treatment of lesions of the cervical spine.)
DePalma, A. F., et al.: Study of the cervical syndrome. Clin. Orthop. 38:135–142, 1965. (A comparison of the results of conservative and surgical treatment for relief of cervical syndromes leads these authors to feel that surgical intervention is justified more often than is generally believed.)
Du Toit, G. T.: The post-traumatic painful neck. Forensic Sci. 3:1–18, 1974. (The common mechanisms of injury, diskography, arteriography, neurologic signs, therapy, fusion, and radiologic investigation are discussed.)
Fielding, J. W., Burstein, A. H., and Frankel, V. H.: The nuchal ligament. Spine 1:3–14, 1976.
Frykholm, R., et al.: On pain sensations produced by stimulation of ventral roots in man. Acta Physiol. Scand. 29:455, 1953.
Gukelberger, M.: The uncomplicated post-traumatic cervical syndrome. Scand. J. Rehabil. Med. 4:140–153, 1972. (Types of trauma, symptoms, and functional radiologic examination are described, and different types of radiologic changes are reported and illustrated.)
Hartman, J. T.: A conversation with J. T. Hartman: The cervical orthosis—does it immobilize? Orthop. Review, Vol. 5, No. 10, 53–57, Oct. 1976.
Holt, E. P.: Fallacy of cervical discography. J.A.M.A. 188:799, 1964.
Jackson, R.: The Cervical Syndrome, ed. 4. Charles C Thomas, Springfield IL, 1971.

Jacobson, E.: Progressive Relaxation, ed. 2. University of Chicago Press, Chicago, 1938.

Krout, R. M. and Anderson, T. P.: Role of anterior cervical muscles in production of neck pain. Arch. Phys. Med. Rehabil. 47:603–611, 1966.

Lalli, J. J.: Cervical vertebral syndromes. J. Am. Osteopath. Assoc. 72:121–128, 1972. (A review of the current medical literature on cervical vertebral syndromes and their treatment is presented along with a useful discussion of the functional anatomy, symptoms, diagnosis, and conservative medical management.)

Lance, J. W.: Headache. In Critchley, M. (ed.): Scientific Foundations of Neurology. F. A. Davis Co., Philadelphia, 1972, pp. 169–175.

Lourie, H., et al.: The syndrome of central cervical soft disk herniation. J.A.M.A. 226:302–305, 1973. (Case histories illustrating the syndrome of central cervical soft disk herniation are presented, and attention is called to possible pitfalls in diagnosis and management.)

MacNab, I.: The whiplash syndrome. Clin. Neurosurg. 20:232–241, 1973. (Doubts about the validity of a diagnosis of "litigation neurosis" precede a discussion of the mechanics of the cervical spine.)

Maigne, R.: Orthopedic Medicine. Charles C Thomas, Springfield, 1972.

McCall, I. W., et al.: The radiologic demonstration of acute lower cervical injury. Clin. Radiol. 24:235–240, 1973. (This paper emphasizes the limitations of anteroposterior and lateral radiographs and describes a simple technique for supine oblique examination of the immobilized patient.)

Miles, W. A.: Discogenic and osteoarthritic disease of the cervical spine. J. Nat. Med. Assoc. 66:300–304, 1974. (A brief review of the anatomy and physiology of the cervical spine is followed by discussion of symptoms, radiographic and pathologic findings, and radiographic assessment of operative intervention.)

Nicholas, G. G.: Initial management of injury to the neck. Pa. Med. 77:47–48, 1974. (Emergency measures are described to provide an adequate airway and to control hemorrhage in patients with neck injury.)

Ostfeld, A. M.: The Common Headache Syndromes: Biochemistry Pathophysiology Therapy. Charles C Thomas, Springfield, IL, 1962.

Rothman, R. H., and Simeone, F. A. (eds.): The Spine, 2 volumes. W. B. Saunders, Philadelphia, 1975.

Sano, K., et al.: Correlative studies of dynamics and pathology in whip-lash and head injuries. Scand. J. Rehabil. Med. 4:47–54, 1972.

Stainsbury, P., and Gibson, J. F: Symptoms of anxiety and tension and accompanying physiological changes in the muscular system. J. Neurol. Neurosurg. Psychiatry 17:216–224, 1954.

Veleanu, C., and Diaconescu, N.: Contribution to the clinical anatomy of the vertebral column consideration on the stability and the instability at the height of the "vertebral unit." Anat. Anz. 137S:287–295, 1975.

Wiberg, G.: Back pain in relation to the nerve supply of the intervertebral disc. Acta Orthop. Scand. 19:211, 1949.

CHAPTER 5

Neurovascular Compression Syndromes

Abnormal compression of the neurovascular bundle containing the subclavian artery and vein and the brachial plexus can cause a specific syndrome in the upper extremity. The syndrome has been termed cervicodorsal outlet (C.D.O.), neurovascular compression, or shoulder girdle syndrome and has been further subdivided into anterior scalene, claviculocostal, or pectorales minor syndromes. All terms implicate a similar medical condition.

The brachial plexus descends from the cervical vertebrae as roots that merge into trunks, into cords, and ultimately form peripheral nerves (Fig. 115). The symptoms of the syndrome have been attributed to vascular compression but careful clinical evaluation implicates pressure upon the median trunks involving eighth cervical to first thoracic roots with symptoms referred to the little and ring fingers.

The subclavian artery arches over the first rib behind the anterior scalene muscle and in front of the middle scalene muscle (Fig. 116). It then proceeds under the clavicle and enters the axilla beneath the smaller pectoral muscle. The subclavian vein courses adjacent to the artery except that it passes anteriorly to the anterior scalene muscle. The entire neurovascular bundle is completely confined within spaces formed by dense fascia.

During arm shoulder movement the scapula moves to 45 degrees of abduction and the glenohumeral joint permits full overhead elevation of the arm. In this movement the axillary artery is bent 180 degrees from its dependent position and is pulleyed across the coracoid process and the head of the humerus.

SYMPTOMS

Symptoms of encroachment of the neurovascular bundle consist of pain in the fingers, hand, forearm, and occasionally shoulder with

142

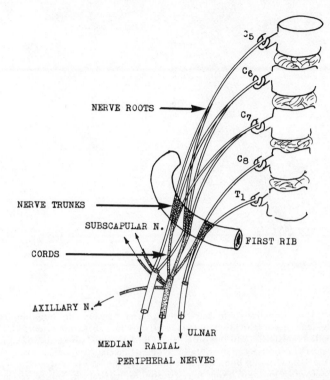

NERVE ROOTS →

C_5

C_6

C_7

C_8

T_1

NERVE TRUNKS →

SUBSCAPULAR N.

CORDS →

FIRST RIB

AXILLARY N.

MEDIAN   RADIAL   ULNAR

PERIPHERAL NERVES

FIGURE 115. Brachial plexus (schematic). The brachial plexus is composed of the anterior primary rami of segments C_5, C_6, C_7, C_8, and T_1. The roots emerge from the intervertebral foramina through the scalene muscles. The roots merge into three trunks in the region of the first rib. The trunks via divisions become cords that divide into the peripheral nerves of the upper extremities.

paresthesias in the eighth cervical and first thoracic dermatome area (Fig. 117). These areas are the posterior aspect of the arm just below the axillary crease with radiation into the palm and two middle fingers. The first thoracic area is the ulnar aspect of the forearm and the little finger.

Numbness may be noticed in the areas of paresthesias. Vascular symptoms may include coldness and discoloration.

CAUSES

The triangle formed by the anterior scalene muscle, the middle scalene muscle, and the first rib through which the subclavian vessels and the brachial plexus emerge remains reasonably constant with no occlusion of its contents. How compression occurs is related to various deviations and abnormalities.

1. Accessory cervical rib with or without fibrous extension may

143

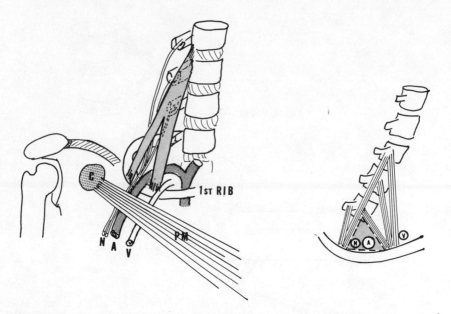

FIGURE 116. The supraclavicular space. The scalene muscles originate from the cervical spine and divide to contain the brachial plexus *(N)* and the subclavian artery *(A)*. The middle scalene muscle is posterior and the anterior scalene muscle is anterior to the artery. The subclavian vein *(V)* is anterior to the anterior scalene muscle. After passing over the first rib (not shown) the neurovascular bundle passes under the smaller pectoral muscle. The clavicle covers the neurovascular bundle and lays parallel to the first rib. The coracoid process is labeled C.

narrow the interscalene triangle and cause angulation of the subclavian artery.

2. The subclavian artery may pass through instead of behind the anterior scalene muscle.

3. Anomalies of the blood vessels or the bony and muscular components may exist.

Since all three abnormalities exist from early life, it is strange that these anomalies become symptomatic. This occurs usually from impairment of posture and musculature causing sagging or drooping of the shoulders which, in turn, exerts tension on the neurovascular bundle.

The scalene muscle raises the first rib and thus aids deep respiration. These muscles also flex and rotate the head. These two basic muscular actions explain the Adson maneuver.

The Adson maneuver tests neurovascular occlusion from *scalene triangle constriction*. The patient (1) hyperextends the neck, (2) turns the head to the side of the area of paresthesia, and (3) takes a deep breath and holds the inspiration. The arm is held at the patient's side and the arterial

144

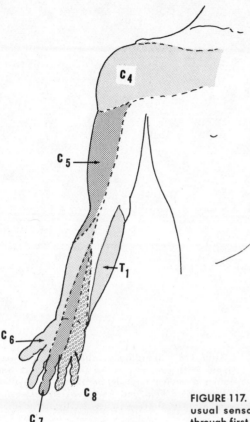

FIGURE 117. Sensory map of the arm. Depicts the usual sensory areas supplied by first cervical through first thoracic roots.

pulse is checked. A positive test reproduces the paresthesia in the hand and occludes the pulse.

Neurovascular compression can be caused by the *costoclavicular syndrome* in which the neurovascular bundle is compressed between the descending clavicle and the first rib. This occurs especially when the shoulders are thrust backward and downward. Again a postural causation is elicited. The maneuver to test this syndrome is to place the patient in an exaggerated military posture with shoulders drawn backward and downward. A simultaneous deep inspiration further confines the costoclavicular space.

A third mechanism of neurovascular compression is the *pectoralis minor syndrome* or the *hyperabduction syndrome*. The compression occurs where the neurovascular bundle passes beneath the tendon of the smaller pectoral muscle. The maneuver for this sign is full elevation of the arm above and behind the head. This arm position causes the bundle to curl

145

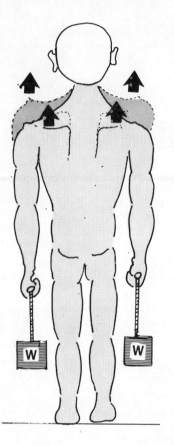

FIGURE 118. Standing scapular elevation exercises. With proper posture (tilted pelvis and flattened cervical lordosis) both arms are rhythmically elevated, held, and slowly lowered. Increasing weights are used. Elbows must be fully extended.

under the coracoid process and be compressed between the tense smaller pectoral muscle and the underlying costal borders. All three mechanical factors may be involved in causing neurovascular compression symptoms.

TREATMENT

Treatment should be conservative and extensive before more drastic diagnostic procedures such as venography, arteriography, or surgical exploration is undertaken. Improvement of posture and change of faulty sitting and walking habits is mandatory.

The fascia of the cervical area and the thoracic inlet through which all the neurovascular bundle elements pass may be thickened and contracted. The first and second ribs may be immobilized and the scalenes restricted in their motion. This fascial limitation may be partially released by cervical traction, but passive stretch of these tissues may be necessary.

With one of the physician's hands positioned against the clavicle and

146

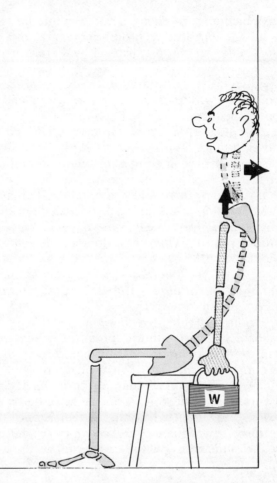

FIGURE 119. Posture-scapular elevation exercises. Patient is seated with back to wall, his head and neck pressed against the wall, which decreases the cervical lordosis. With arms fully extended and dependent, weights are lifted in a shrugging motion. Weights vary from 5 to 30 pounds.

the first ribs, the head is rotated and laterally bent away from the symptomatic side and these tissues stretched in a gentle but persistent intensity. Gradual increase in range can be appreciated. Full neck range of motion must also be regained.

The usual treatment program, outlined in Chapter 4, which includes traction, exercise, and posture training, pertains in this syndrome also but with additional aspects.

Because loss of muscle tone and strength of the shoulder girdle play an instrumental role in the syndrome, attention has to be directed to this deficiency. The muscle weakness of middle aged disuse atrophy, the

147

pseudodystrophic muscular changes noted in middle aged menopausal women, the muscle tone loss in chronic depression, or a complication of debility associated with chronic diseases—all are considered causative factors.

The individual with a rounded shoulder posture with adaptive shortening of muscle groups, such as pectoral group and neck extensors, is more prone to neurovascular compression.

Resistive exercises to shoulder elevators (Fig. 118) are desired for both strength and endurance. With weights held by arms in the dependent position these weights are lifted in a shrugging movement. During this exercise, the low back lordosis must be decreased, i.e., the pelvis tucked in. Strength is gained by increasing the weight and endurance gained by prolonging the duration of sustaining the elevated position.

During this scapulae elevating exercise the neck must also be kept in a flattened lordotic posture. This can be done in a seated position against a walled surface (Fig. 119).

Neck flexor strength is necessary here and the flexors should be strengthened by lifting the head in the supine position with the chin held in a downward position towards the sternum.

Every day activities should be reviewed to avoid faulty sitting, walking, and driving habits. Overcoming depression pharmaceutically or by counceling should also be considered early.

There are numerous other causes that will not respond to conservative measures. These include clavicular fractures healed with excessive callous, post-irradiation fibrosis, gross thoracic deformity, metastatic plexus invasion, and vascular thrombosis. Persistence of symptoms in spite of an otherwise adequate conservative treatment program should always alert the clinician to pursue further diagnostic or even surgical intervention.

BIBLIOGRAPHY

Adson, A. W.: Cervical ribs: Symptoms, differential diagnosis for section of the insertion of the scalenus anticus muscle. J. Internat. Coll. Surgeons 16:546, 1951.

Beyer, J. A., and Wright, I. S.: Hyperabduction syndrome, with special reference to its relationship to Raynaud's syndrome. Circulation 4:161, 1951.

Cailliet, R.: Neck and Arm Pain. F. A. Davis Co., Philadelphia, 1964.

Dejerine, J.: Semiologie des Affections du Systeme Nerveux. Masson et Cie, Paris, 1926.

Edgar, M. A., and Nundy, S.: Innervation of the spinal dura mater. J. Neurol. Neurosurg. Psychiatry 29:530, 1966.

Lord, J. W., and Rosati, L. M.: Neurovascular compression syndromes of the upper extremity. Clin. Symp. 10:1958.

Naffziger, H. C., and Grant, W. T.: Neuritis of the brachial plexus mechanical in origin: The scalenus syndrome. Surg. Gynecol. Obstet. 67:722, 1938.

Nelson, P. A.: Treatment of patients with cervico-dorsal outlet syndrome. J.A.M.A. 27:1575, 1957.

Telford, E. D., and Mottershead, S.: Pressure at the cervicobrachial junction. J. Bone Joint Surg. 30[Br]:249, 1948.

CHAPTER 6

Shoulder Pain

Although there are numerous joints forming the shoulder complex, the joint causing the greatest amount of pain and disability is the glenohumeral joint. This joint is an incongruous joint that permits greater mobility at the expense of stability.

FUNCTIONAL ANATOMY

Congruent joints have symmetric opposing articular surfaces and a central axis of rotation; thus they have a capsule that permits movement equally in many directions and muscles that rotate the part about the central axis (Fig. 120). Incongruent joints have asymmetric articular surfaces; therefore they have a moving axis of rotation, a redundant capsule, and muscular action that must simultaneously stabilize the joint and move it in a gliding motion.

The glenohumeral joint is one of these incongruous joints. This incongruity is the basis of much discomfort, pain, and disability. Pain can result from faulty use, trauma, or degenerative changes. Full evaluation of the history and examination of the active and passive range of motion usually can reveal the pathologic factors and lead to a realistic diagnosis and meaningful treatment.

Abduction of the glenohumeral joint requires several movement components.

1. The musculature must act upon the incongruous joint. The deltoid muscle abducts the arm only if there is slight abduction of the humerus; otherwise deltoid action acts along the shaft and merely elevates the head of the humerus against the overhanging acromion and coracoacromial ligament (Fig. 121). The cuff muscles must secure the head of the humerus against the glenoid fossa, must rotate the humeral head to abduct the arm, and must begin depressing the humeral head down along the glenoid fossa (Fig. 122). This last motion allows the greater

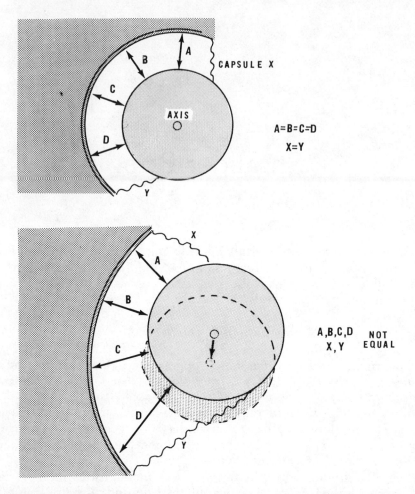

FIGURE 120. Congruous and incongruous joints. *Above,* Congruous joint with parallel surfaces of the socket and the articulating head. The axis of rotation is central and the head rotates about this axis. The capsule has general flexibility and muscle action is equal about the axis merely moving the head about the axis. The symmetric surfaces with the head deep seated gives the joint stability.

Below, The incongruous joints have asymmetric surfaces with the capsule concavity much broader than the convex head. Joint space differs throughout the articulation. The axis of rotation varies with the position of the head. Muscular action must fix the head and imitate the gliding motion. The capsule must permit this type of motion.

tuberosity to pass under the overhanging acromion during the abduction arc of 90 degrees. The cuff muscles also rotate the humerus about its longitudinal axis to externally or internally rotate the arm. This simultaneous rotation to the movement of abduction also allows the greater tuberosity to pass unrestricted behind and under the acromion.

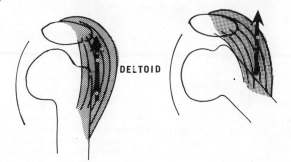

DELTOID

FIGURE 121.   Deltoid function. *Left*, With the arm dependent, the deltoid acts along the longitudinal axis of the humerus elevation of the arm. *Right*, With some abduction initiated by the cuff, the deltoid pulls at an angle to the humerus, becoming the prime mover of abduction.

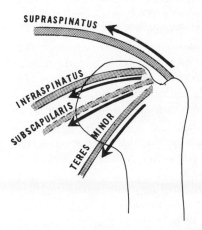

SUPRASPINATUS

INFRASPINATUS

SUBSCAPULARIS

TERES MINOR

FIGURE 122.   Rotator cuff mechanism. The supraspinous (supraspinatus) muscle pulls the head of the humerus into the glenoid and slightly rotates the humerus into abduction. The infraspinous (infraspinatus) muscle also rotates the head and slightly pulls it down. The teres minor muscle pulls in a more downward direction. The subscapular (subscapularis) muscle pulls the head into the glenoid, but its main rotatory action is to internally rotate the humerus about its longitudinal axis.

2. The cuff must be intact and without any thickening from inflammation, calcium, or a scar to allow passage through the narrow suprahumeral space between the humeral head, the acromion, and the coracohumeral ligament.

3. The subdeltoid bursa, which facilitates movement between the cuff and the adjacent undersurface of the deltoid and the subacromial bursa, must be normal and not be inflamed, swollen, or calcified.

4. The capsule must have sufficient redundancy to permit the humeral head to glide down the glenoid fossa and rotate in its two physiologic directions (Fig. 123).

5. Nerve supply to the pertinent muscles must be intact.

6. The articulating surfaces of the glenoid fossa and the humerus must be smooth, thus free of significant osteoarthritis, and the capsular synovial fluid must have the normal lubricating consistency, i.e., clear, viscid, and not infectious nor hemorrhagic.

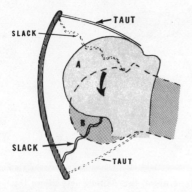

FIGURE 123. Glenohumeral capsule. With arm dependent *(solid line)* the superior capsule is taut and supports the arm. The inferior capsule is slack, and thus, permits the incongruous movement *(dotted lines)* of downward gliding during arm abduction.

Since the glenohumeral joint will abduct only 90 degrees with the arm in direct sagittal plane before the tuberosity impinges upon the overhanging acromion, the scapula must simultaneously and proportionately rotate upon the thoracic cage. By this simultaneous scapulohumeral rhythmic motion, the acromion moves ahead of the greater tuberosity, thus permitting further abduction and overhead arm elevation. The deltoid muscle is also kept at a more physiologic length for optimum contraction as the scapula rotates (Fig. 124).

The internally rotated humerus abducts only to 60 degrees and the externally rotated humerus to 120 degrees. This difference is due to the rotation about the humeral axis, which passes the greater tuberosity behind the overhanging acromion. The rotator cuff muscles are responsible for this rotation as well as the movement of abduction. The motions must occur simultaneously to prevent entrapment of the cuff between the overhanging acromion and the greater tuberosity of the humerus.

Dorsal kyphotic posture places the overhanging acromion in a forward and downward position, which obstructs humeral abduction and elevation. This latter fact accounts for the limitation of shoulder range of motion in a round-back posture such as is noted in poorly postured individuals, kyphosis of the aged, or the posture of the depressed individual.

The circulation of the cuff plays an important part of the attrition and degeneration noted in all individuals as they approach their fourth or fifth decade. The "critical" zone of the cuff is highly vascularized by anastomosis of the bone vessels and the muscular branches (Fig. 125). It is, however, relatively ischemic when the dependent hanging arm causes occlusion by traction and the actively abducting elevating arm occludes the blood vessels from contraction of the cuff muscles. Theoretically the only time that circulation is adequate is in the supported inactive arm.

152

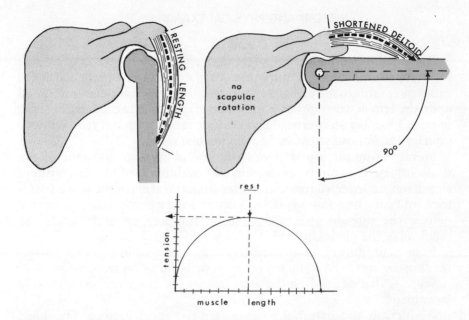

FIGURE 124. Maximum muscle tension related to muscle length. Greatest muscle tension (strength) is with the muscle approximately 10 percent longer than resting length. The deltoid is shown at resting length and in abduction, without scapular rotation, gradually loses efficient tension. This reveals need for simultaneous scapular rotation to maintain optimum deltoid strength.

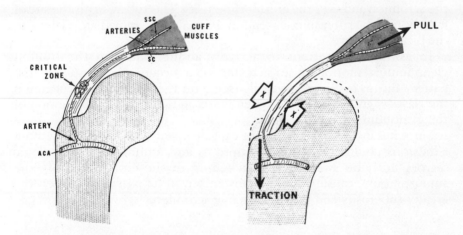

FIGURE 125. Blood circulation. *Left,* Circulation to the rotator cuff. The arterial branch from the anterior circumflex artery *(ACA)* enters from the bone. The suprascapular *(SSC)* and the subscapular *(SC)* branches merge to enter from the muscle. The critical zone of the tendon is an anastomosis which is patent when the arm is supported and inactive. *Right,* Traction upon the cuff from the dependent arm or from pull of the contracting cuff muscle elongates the tendon and renders the critical zone *(arrows)* relatively ischemic.

153

Pain and limitation of range of motion usually are the result of a direct injury or injury through faulty use. A fall on the outstretched arm can cause direct soft tissue damage to the glenohumeral joint. Force imposed upon the arm in which physiologic range of motion is exceeded can result in pain. Opening an overhead garage door can wrench the arm. Athletic injuries can impose stress far beyond normal range.

Merely using the arm and violating the physiologic movements can cause injury. This can be exemplified by adduction of the arm without simultaneous external rotation or elevating the arm overhead while in a forward flexed or dorsal kyphotic posture. Injury occurs that may stretch or tear the capsule or stretch, tear, or impinge upon the cuff. The cartilage of the glenohumeral joint may also sustain damage.

In an acute injury, examination reveals a patient in obvious distress holding the arm in an adducted position with the elbow flexed. The usual posture is that of holding the arm against the body and avoiding any movement.

Abduction and external rotation are the most limited. Overhead elevation is impossible and frequently internal rotation with posterior flexion (such as reaching behind into a rear pocket or fastening a bra) is also limited. Passive attempt at these ranges are equally restricted to the active limitation.

Tenderness can frequently be elicited anteriorly immediately below the acromion and over the greater tuberosity. This tuberosity is the site of insertion of the supraspinous tendon. It is palpable immediately lateral to the bicipital groove.

In asking the patient to actively abduct the arm the limited glenohumeral motion causes all abduction to occur at the scapula and the patient shrugs his shoulder girdle (see 5 on Fig. 133). This shrugging is the classic sign of shoulder peritendinitis and depicts limited motion of the glenohumeral joint.

Since it is inconceivable that the capsule inflammation with or without adhesions could cause these symptoms and findings, the pathologic factors must be swelling or thickening of the cuff (especially the supraspinous), causing it to be obstructed in its passage between the greater tuberosity and the overhanging acromion.

TREATMENT

Treatment must be based on the pathomechanics that cause the symptoms and the findings.

Physical Therapy

Heat is usually not indicated because its use may cause increased symptoms and findings. Conceivably, heat increases swelling of the inflamed cuff, thus causing greater encroachment within the normally narrow suprahumeral space.

Ice is favorable. Ice causes local anesthesia and decreases pain; it causes vasoconstriction and decreases swelling; and it acts on the muscle spindle system which decreases the spasm of the periarticular muscles.

Immobilization

As in most acute traumatic inflammatory reactions the part must be rested. Movement must be restricted but only for a specified time. Prolonged immobilization must be avoided as muscular contracture results, the capsule thickens and becomes adherent, muscle atrophy results, and habit pattern of the "frozen shoulder" may occur.

Immobilization is best accomplished by a sling that holds the arm against the body with slight internal rotation and suspends the arm against gravity. Motion, however, must be started within 24 to 48 hours to prevent contracture and atrophy.

Medication

Pain medication is valuable. Usually medication of the anti-inflammatory type (Butazolidin, Butazolidin Alka, and Indocin) is preferable for a brief oral course if no contraindication exists.

Exercise

To maintain range of motion or regain range during the acute phase of pain and restriction, passive movement utilizing gravity traction and avoiding muscular contraction is preferable. This activity is best accomplished by properly executed pendular exercises (Fig. 126).

With the upper back flexed forward to approximately 90 degrees, knees slightly bent to avoid low back strain, and optimal support against a chair or desk with the other arm, the involved arm is dangled. Movement of the body passively swings the dependent arm in an anteroposterior direction, in a lateral direction, and in a rotatory direction. The arm must be swung passively by moving the body, thus avoiding any muscular effort upon the glenohumeral joint.

It would appear that holding a weight in the hand would aid by increasing the swing momentum or by increasing traction force. This is not so because the muscular effect of the hand gripping the weight would

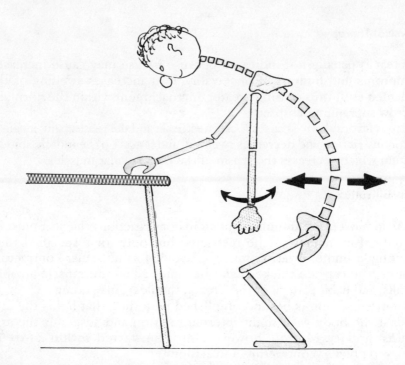

FIGURE 126. Pendular exercise. The patient bends forward flexing the trunk to right angles. The involved arm is dangled without muscular activity of the glenohumeral joint. The body actively sways thus passively swinging the dependent arm in forward flexion-extension, lateral swing, and rotation. The body can be supported by placing the other arm upon a table or chair.

cause proximal muscular contraction of the shoulder muscles, a condition which is not desired. A weighted bracelet of 1 or 2 pounds fastened about the wrist by velcro is more effective. Exercises should be done hourly during waking hours.

As the range of motion increases, active exercises, avoiding abduction, may begin. Internal and external rotation (Fig. 127) can be accomplished by assuming the seated position with the upper arm supported on a table and the 90 degree flexed elbow moving alternately between internal and external rotation. By leaning forward against the forward stretched arm further elevation of the arm can be gained.

Ultimately an overhead pulley can gradually achieve full overhead elevation of the arm and movement of the arm posteriorly towards the posterolateral aspect of the head. The other arm/hand supplies the force to accomplish this range of motion (Fig. 128).

Push-ups towards and away from a wall corner with the hands against each wall and gradually moving up the wall to and above shoulder height will increase external rotation range.

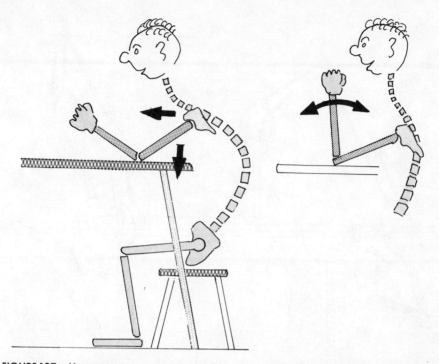

FIGURE 127. Home exercise to increase shoulder range of motion. Seated with arm supported upon table, the patient moves forward and downward to increase range of arm towards elevation. The forearm, bent at right angles, internally and externally rotates, thus further increasing range. A weight in this exercise can be held in the hand.

Posterior flexion can be increased by placing both hands on a table or cabinet behind the body and gradually doing increasing deep knee bends (Fig. 129).

Wall fingerclimbing is effective if the climbing is done against a wall faced by the patient. Wall climbing in abduction causes more scapular shrugging with little or no glenohumeral movement and should be avoided.

Injection

Intra-articular or suprahumeral injection early in the acute phase is not indicated. Any amount of injected material further encroaches upon the already overcrowded suprahumeral space.

A painful shoulder frequently can be benefited by a suprascapular nerve block which is simply performed and can give relief. The suprascapular nerve enters the supraspinatus area through the scapular notch that is situated approximately ¾ of the distance of the scapular

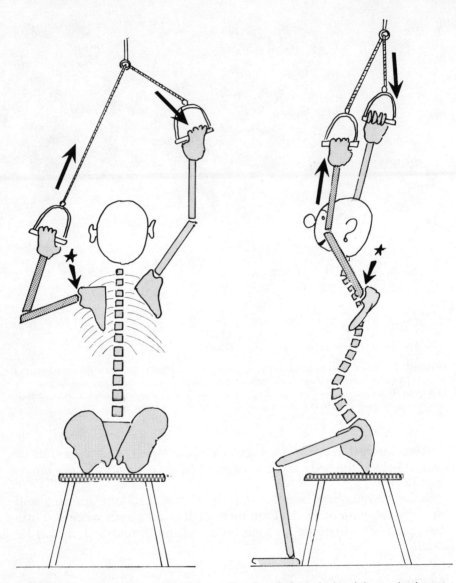

FIGURE 128. Overhead exercise. With a pulley placed above the head the involved arm is passively elevated by the normal arm. By having the pulley slightly behind the head the arm gets further range of motion to overcome one of the subtle signs of limitation.

spine measured from the superior medial angle (Fig. 130). As the needle approaches the notch the local anesthetic is diffused through the soft tissues. Since the nerve is accompanied by blood vessels through the notch, aspiration before injection of the anesthetic agent is mandatory.

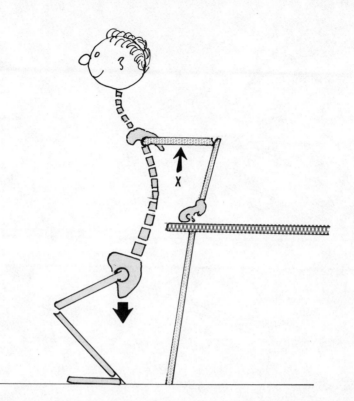

FIGURE 129. Exercise to stretch anterior capsule and increase posterior flexion. The patient places both hands on a table behind the back and performs gentle deep knee bends. This elevates arms *(small arrow)* and increases posterior flexor range of motion.

After 24 to 48 hours, suprahumeral injection of a steroid and anesthetic such as Xylocaine or Novocain is beneficial. To be effective the injection must be given in the suprahumeral joint in the region of the supraspinatus tendon insertion (Fig. 131). This site can be clinically ascertained by following the clavicle laterally until the acromion is reached. The inferior margin of the acromion overlaps the suprahumeral sulcus. The humerus can be palpated immediately inferior to the acromial process and the bicipital groove located. Verification of the bicipital groove is gained by actively or passively rotating the humerus about its longitudinal axis. The groove slips past the palpating finger. The injection needle enters just superior and lateral to the bicipital groove, which is the site of the greater tuberosity and the insertion of the supraspinatus tendon. The needle shortly after skin penetration is inserted in an upward posterior direction under the acromion. No bone contact should occur. Injections are usually immediately effective and can be repeated for two to three injections at weekly intervals.

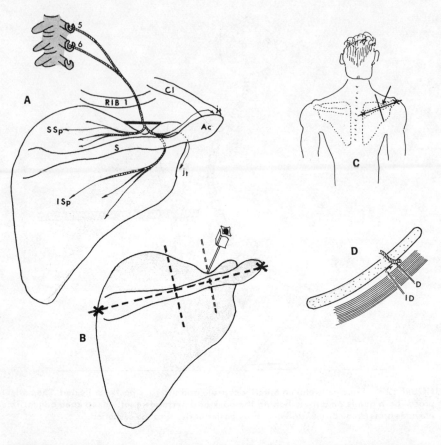

FIGURE 130. Site of suprascapular nerve block. *A,* Normal anatomy of suprascapular nerve. *B,* Surface anatomy for site of injection: Measure spine of scapula and estimate midpoint. Nerve is halfway between midpoint and lateral tip of spine *(C).* Direct needle entrance to the nerve *(D). ID* shows indirect approach of anesthetic agent to nerve when needle is slightly distant to nerve emergence.

Active Assisted Motion

Active assisted motion by a therapist or a trained member of the family may be necessary in shoulders that have been immobile for a period of time or when pain has been severe, thus preventing motion. Increased joint "play" treatment, considered to be gentle manipulation, increases lateral and anteroposterior motion which is not possible done actively by the patient. This movement is done gently but with some force to stretch the capsule. Traction along the longitudinal axis of the humerus can also be exercised.

Active motion in all directions assisted by the therapist can increase range. This is termed "rhythmic stabilization" in that as the therapist

160

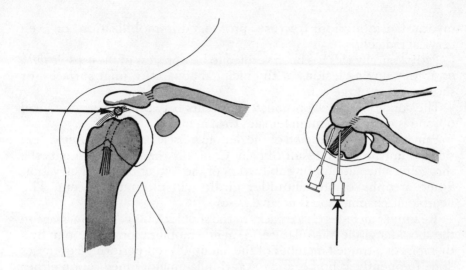

FIGURE 131. Site of injection in pericapsulitis. *Left,* Region of supraspinatus insertion in the suprahumeral space. The region is palpable immediately below the overhanging acromion and by palpating the greater tuberosity just lateral to the bicipital groove of the humerus. *Right,* Insertion of needle viewed from above. Two directions of entrance are shown, with the arrow depicting that shown in the anterior view.

attempts a movement (e.g., forward flexion) this is immediately resisted by the patient. The movement shifts immediately in the opposite direction (e.g., backward flexion) which is resisted by the patient. Alternating this exercise in a rhythmic pattern the joint does not actively move, but the joint is mobilized with increased capsular flexibility. By active muscular contraction of an isometric type, muscle weakness and atropy is minimized or corrected.

Posture

Posture is very important in improving shoulder motion. The rounded-shoulder posture, excessive dorsal kyphosis, physiologically decreases range of motion and adds further entrapment of the supraspinatus cuff by lowering the overhanging acromion. All the posture exercises described previously should be employed in complete treatment of the shoulder.

THE FROZEN SHOULDER

The frozen shoulder, referred to as adhesive capsulitis or periarthritis, denotes marked limitation both actively or passively of all motion of the glenohumeral joint. This is a clinical picture that may follow tendinitis,

161

myocardial infarction, fracture, prolonged immobilization, or even cervical radiculitis.

Pathologically this has been verified to be a process of the capsule *with no* accompanying lesions of the bicipital apparatus, joint surfaces, or short rotation of the cuff.

The condition usually is unilateral, occurs between the age of 45 and 60, and is more frequent in females than in males.

Pain is noted in the region of the deltoid muscle, motion is limited, and sleep is impaired as a result of pain. Clinically the patient presents the shrugging when attempting abduction of the arm or overhead elevation. Some atrophy of the shoulder girdle will ultimately occur. The neurologic examination is normal.

Treatment includes (1) ice packs to the shoulder, (2) general massage to the shoulder girdle musculature, (3) gentle mobilization of the joint by a therapist or a trained member of the family, (4) self-performed exercises done frequently at home, and (5) anti-inflammatory medication either orally or by injection. This condition is chronic and will require long periods of treatment.

THE SUBTLE SIGNS OF RESIDUAL PERICAPSULITIS

Many patients continue to have symptoms in spite of their feeling that the shoulder has good range of motion and in spite of the fact that a cursory examination frequently fails to reveal minor residuals of joint range restriction. These minor limitations can be considered subtle signs and explain persistence of symptoms.

Figure 132 reveals normal active range of motion. Figures 133 and 134 reveal subtle signs of shoulder limitation—signs of which the patient may not be aware and signs which the physician will miss if he does not give a complete examination of range of motion. When the patient abducts the arm, slight scapular elevation (shrugging) may be noted as compared to the normal side (Fig. 133-5). Overhead elevation of the arm may fail to reach the head to the same extent as the normal and, when viewed from the side, the arm is not extended as posteriorly as the normal arm (Figs. 133-6 and 134-9). When the patient reaches posteriorly, the involved arm does not reach as high between the shoulder blades (Fig. 134-11). With arms at side and elbows bent 90 degrees, external rotation of the painful shoulder is not as complete as the normal side (Figs. 133-7, 134-8, and 134-10).

ROTATOR CUFF TEAR

Injuries to the shoulder can disrupt the cuff with either a complete or partial cuff tear (Fig. 135). A complete tear may clinically resemble an

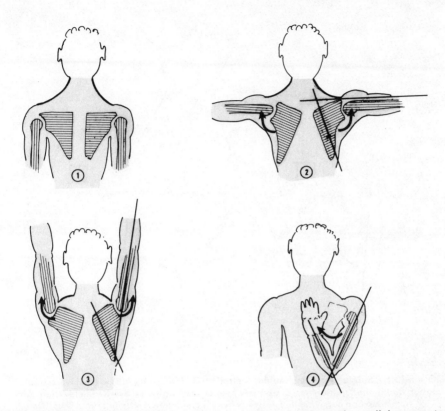

FIGURE 132. Normal scapulohumeral motions. *1*, Normal stance with parallel symmetric shoulder girdles; *2*, Symmetric abduction with equal proportional glenohumeral abduction and scapular rotation; *3*, Symmetric full overhead arm elevation; *4*, Posterior arm flexion and internal rotation.

acute peritendinitis in that there is pain, marked limitation, and tenderness. The arm cannot be abducted at the glenohumeral joint and the patient shrugs. Once the pain is decreased, as it can be by an anesthetic suprahumeral injection (performed as previously indicated), although the arm cannot be actively abducted it can be passively abducted and then actively held by the patient in the acquired position.

The mechanism that explains this phenomenon is that the cuff muscles cannot seat the humeral head firmly into the glenoid fossa nor rotate the humeral head to place the deltoid in a functional position (of humeral abduction). Passive abduction by the examiner substitutes for cuff action and places the humerus in a sufficiently abducted position to allow the deltoid to become a prime mover. Any force applied to the abducted arm overwhelms the sustained position because the humerus is not held firmly into the glenoid fossa.

A partial tear reacts exactly as does peritendinitis with the torn fibers

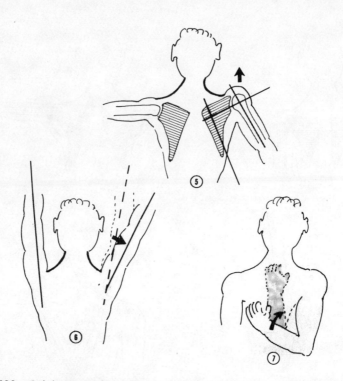

FIGURE 133. Subtle signs of shoulder limitation. 5, Shrugging with excessive scapular rotation and limited glenohumeral abduction; 6, Limited right arm overhead elevation. Arm "away" from head and ear; 7, Limited posterior flexion and internal rotation. Hand fails to reach normal interscapular distance of reach.

contracting and forming a swelling of the cuff, which obstructs free motion in the suprahumeral space.

Treatment

Treatment of the partial cuff tear is similar to that of the tendinitis other than that active and resisted exercises are done more carefully and gently to avoid further tearing by the applied forces. Only violent or unusual force can convert a partial tear into a complete tear.

Surgical repair should be considered in a complete tear in a reasonably young person whose activities and profession require full range of shoulder motion with good strength. However, in elderly or severely debilitated patients, surgical repair may not be successful or lasting. With full understanding by the patient of the possible outcome of surgery, every patient, nevertheless, can be considered a surgical candidate. Postoperative care will require a full exercise program as outlined for the other shoulder conditions.

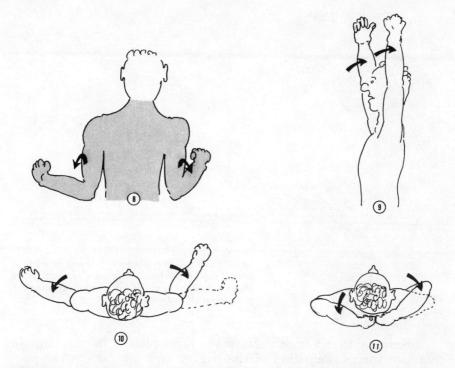

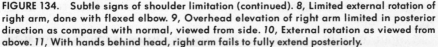

FIGURE 134. Subtle signs of shoulder limitation (continued). *8*, Limited external rotation of right arm, done with flexed elbow. *9*, Overhead elevation of right arm limited in posterior direction as compared with normal, viewed from side. *10*, External rotation as viewed from above. *11*, With hands behind head, right arm fails to fully extend posteriorly.

ACROMIOCLAVICULAR PAIN

Patients with complaints of shoulder pain may have the site of their pain at the acromioclavicular joint. Diagnosis can be reasonably simple if this possibility is considered in the differential diagnosis. The site of pain and tenderness is at the acromioclavicular joint. Shoulder motion may be restricted, but the pain and crepitation definitely is at the acromioclavicular joint. Mere scapular elevation without glenohumeral motion also aggravates or initiates the pain. An injection of a local anesthetic into the joint is both diagnostic and therapeutic. Local steroid and anesthetic injections are frequently beneficial with no other therapy necessary. Surgical intervention should be sought if pain and disability defies conservative treatment and the symptomatology is of significant severity.

Pain may be referred for extra shoulder sources such as

1. Referred pain from the cervical spine. Reproduction of the pain from cervical spine movement, especially extension with simultaneous lateral motion and rotation to the site of the pain, is pathognomic.

165

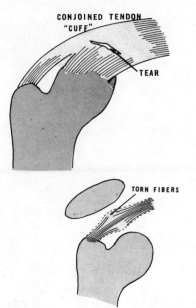

CONJOINED TENDON "CUFF"

TEAR

TORN FIBERS

FIGURE 135. Cuff tear. *Above*, Site and direction of partial cuff tear. *Below*, Retraction of torn cuff fibers forms a thickening of the cuff, thus resembling the thickening of tendinitis.

2. Referred from a trigger zone in the supraspinatus, infraspinatus, or trapezius forces. Palpation of the trigger area reproduces the pain. Injection of the trigger area with a local anesthetic may relieve the referred pain.

3. Pain may be referred from visceral disease such as myocardial ischemia, gallbladder disease, or subdiaphragmatic disease. A careful history, physical examination, and appropriate laboratory or x-ray studies are valuable.

REFERRED PAIN

A frequent complication of the shoulder pericapsulitis/peritendinitis is the hand-shoulder syndrome. The failure of freedom of the glenohumeral joint places the hand in a constant dependent position. It also prevents the hand from being placed in a functional position, and so the hand does not function adequately. The hand swells, further restricting full flexion extension, and more edema results. The edema creates further ischemia and the hand becomes painful, limited, and ultimately useless. Prevention of this complication is the treatment of choice with mandatory finger-wrist exercises in as elevated a position as possible. Mechanical vasoconstriction such as Jobst pneumatic compression is valuable.

If the hand is painful and demonstrates vasomotor changes such as

166

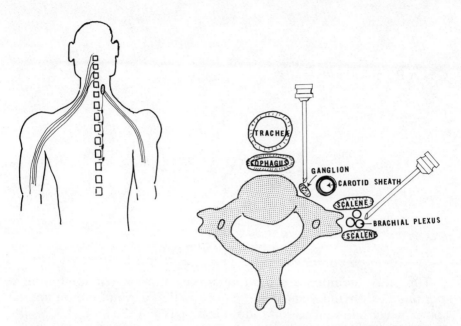

FIGURE 136. Stellate block: clinical sympathectomy. The block is performed from the anterior neck approach, slightly lateral to the trachea with the needle penetrating to the vertebral body. Usually 8 ml. of 1 percent Procaine is effective within 10 minutes.

erythema, blanching, excessive perspiration, coldness, or rubor, a stellate sympathetic chemical nerve block should be given and repeated weekly for two to four injections (Fig. 136). Occasionally when chemical sympathectomy is successful but of brief duration, a surgical sympathectomy should be considered.

When there are no contraindications, a brief course of steroids in large doses (e.g., Prednisone 60 mgm., 48, 36, 24, 12, 8, and 6.6 mgm.) has proven to be of value.

BIBLIOGRAPHY

Cailliet, R.: Shoulder Pain. F. A. Davis Co., Philadelphia, 1966.
Mennell, J. M.: Joint Pain. Little, Brown & Co., Boston, 1964.
Rubin, D.: An exercise program for shoulder disability. Calif. Med. 106:39–43, 1967.
Welfling, J.: Painful Shoulder—Frozen Shoulder. Falio Rheumatologia, Geigy Pharmaceuticals, New York, 1969.

CHAPTER 7

Elbow Pain

FUNCTIONAL ANATOMY

The elbow comprises three joints: the humeroulnar joint, which permits flexion and extension, and the radioulnar and radiohumeral joints, which allow pronation and supination (Fig. 137).

Muscles act upon the elbow in the following manner:

1. The brachial (brachialis), which originates from the lower aspect of the lower half of the humerus and attaches upon the anterior aspect of the coronoid process of the ulna, is the main flexor of the elbow.

2. The long and short heads of the biceps, which unite about the middle of the arm and insert on the medial aspect of the radius, flex and powerfully supinate the forearm.

3. The triceps, which originates from the lower posterior aspect of the humerus (humeral heads) and inserts upon the ulna, extends the elbow.

Examination of the elbow must include both active and passive range of motion and must test flexion-extension (humeroulnar joint) (Fig. 138) and pronation-supination (radioulnar joint) (Fig. 139). Passive joint examination reveals the integrity of both joints including the flexibility of the periarticular tissues. Active range of motion with or without resistance evaluates the muscular function in action, strength, and endurance. Active flexion also reveals the adequacy of innervation to the muscles of the joint, and so impairment can localize lesions of the peripheral nerve, brachial plexus, or cord root level.

The anterior aspect of the joint, the antecubital fossa, contains the biceps tendon, the radial and brachial arteries, the median and ulnar nerves, and the origin of many forearm muscles (Fig. 140). All these tissues are available to examination.

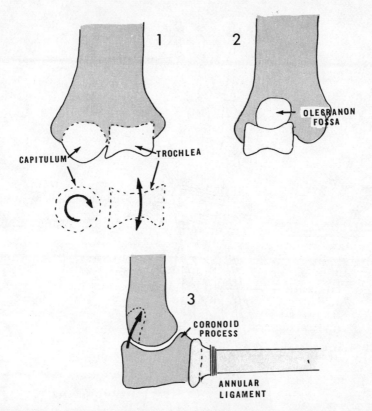

FIGURE 137. Elbow joint bony anatomy. 1, Anterior view depicting the round sphere of the capitulum, upon which the radius rotates, and the spool-shaped trochlea, about which the ulna flexes and extends. 2, Posterior view of the humerus showing the olecranon fossa into which the posterior (olecranon) portion of the radius enters upon elbow extension. 3, Lateral view of the elbow joint.

TRAUMA

Trauma to the elbow usually impairs the humeroulnar joint with resultant limitation of flexion and extension. Rotation may be normal in spite of severe limitation of flexion-extension. If pronation-supination is impaired after trauma, the head of the radius must be suspected to have sustained a fracture.

The presence of traumatic myositis of the brachial muscle in elbow injuries may frequently be overlooked. The following help in differential diagnosis:

1. In injury to the elbow without associated myositis, resisted elbow flexion does *not* cause pain in spite of limited flexion.

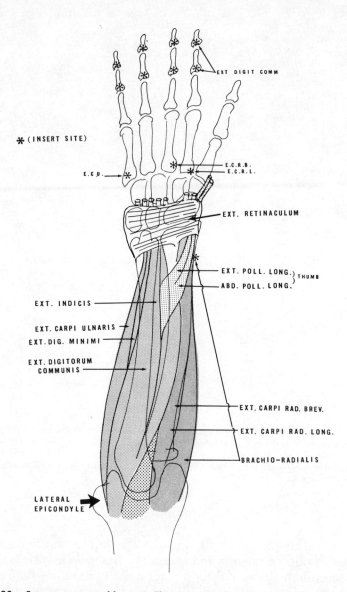

FIGURE 138. Extensor aspect of forearm. The extensor group with particular emphasis on the origin of the common extensor group from the lateral epicondyle. Muscles with asterisks are innervated by the radial nerve.

2. With injury to the brachial muscle, flexion is limited *and painful* at the extreme of flexion. This is the point where the inflamed brachial is squeezed between the humerus and the ulna.

3. When the brachial muscle is injured, resisted extension at midrange causes pain in the region of the brachial by increasing tension upon the

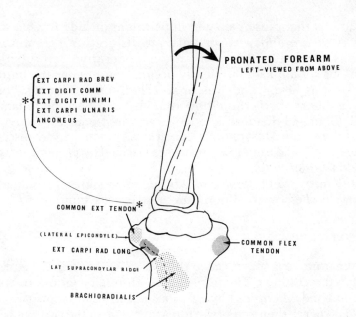

FIGURE 139. Bony aspect of elbow in pronated position. Site of muscular attachment of forearm muscles.

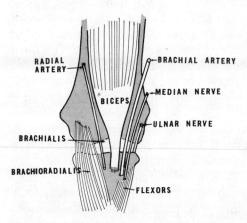

FIGURE 140. Contents of cubital fossa.

inflamed brachial. This does not necessarily occur when there is joint trauma and no associated brachial myositis.

Cyriax[1] places importance on this latter brachial complication, implying that immobilization of the injured elbow, especially with brachial myositis, should be in full flexion and that *no* massage, exercise, or attempt to increase extension range should be permitted.

171

Conditions that cause pain and impairment include trauma with or without fracture and with or without dislocation or subluxation, infectious or rheumatoid arthritis, degenerative arthropathy, or ligamentous sprain. Pain frequently occurs from trauma to contiguous tissues such as in epicondylitis or so-called tennis elbow. There are numerous bursa about the elbow that also can become inflamed or infected. Gout and rheumatoid nodules also are frequently found in the region of the elbow. Most of these conditions enumerated are discernible from the history and examination and are clarified by proper roentgenography and laboratory tests.

Because many major nerves are exposed to pressure and trauma about the elbow, local elbow conditions can cause distal symptoms.

Ulnar Nerve

The ulnar nerve is superficial in the olecranon fossa and, therefore, is subject to direct injury. The nerve passes through a groove behind the medial epicondyle covered by a fibrous sheath that forms the cubital tunnel. The nerve then enters the forearm between the two heads of the ulnar flexor muscle.

The ulnar nerve supplies the ulnar flexor muscle of the wrist and the deep flexor muscle of the fingers and enters the hand to supply the muscles of the hypothenar eminence and the third and fourth lumbricals, all the interosseous muscles, the adductor muscle of the thumb, and the deep head of the short flexor muscle of the thumb (Fig. 141). The ulnar nerve also supplies the dermatome area of the ulnar aspect of the hand—the litle finger, and the ulnar side of the fourth (ring) finger.

The roof of the cubital tunnel is the arcuate ligament, which is taut at 90 degrees of elbow flexion and slack on extension. The floor is composed of the medial ligament and the tip of the trochlea (Fig. 142). The medial ligament bulges in elbow flexion and can compress the nerve. Therefore, prolonged periods of *extreme* flexion should be avoided.

With the arm abducted, such as position of the arm during intravenous injections, full supination of the forearm pulls the tunnel away from pressure. Pronation is more conducive to pressure. Since the sensory fibers of the ulnar nerve are more superficial than the motor fibers, sensory symptoms are more prevalent. Patients with diabetes, alcoholism, and malignancy are more prone to pressure neuropathy.

Ulnar nerve paresis from compression is difficult to document. Without standardization, progression of a lesion or improvement from treatment is difficult to document. An accepted gradation of palsy has been offered by McGowen:

Grade I   Paresthesia and minor hypoesthesia

172

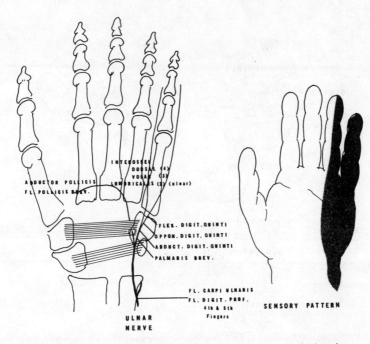

FIGURE 141. The motor and sensory distribution of the ulnar nerve in the hand.

Grade II   Weakness and wasting of interosseous muscles with incomplete hypoesthesia

Grade III   Paralysis of interosseous muscles and severe atrophy of the hypothenar muscles and adductor muscle of the thumb (cause clawing of the ring and little finger)

Compression can also be classified as (1) acute, resulting from a single episode of trauma, (2) subacute or external pressure over a prolonged period, or (3) intrinsic pressure such as osteoarthritic spurs, rheumatoid nodules, ganglia, or soft tissue tumors.

TREATMENT. Conservative treatment of ulnar nerve palsy from external pressure is usually effective. This includes soft pads over the fossa, avoidance of pressure in activities such as writing, and avoidance of excessive flexion. The patient who experiences pressure during sleep may benefit from wearing a light splint to prevent abnormal arm position during sleep. Surgical transplantation, once frequently performed, is no longer favored. Electrophysiologic studies by Payan[2] confirmed sensory recovery to be as rapid and complete from conservative treatment as from surgical transposition.

Examination of the patient with paresthesia of the little and ring fingers must rule out these symptoms as originating from cervical

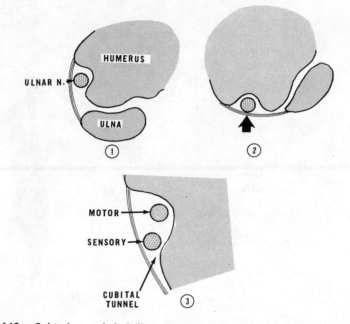

FIGURE 142. Cubital tunnel. *1,* Full supination removes the ulnar nerve from external pressure. *2,* Full pronation places the nerve in position of pressure. Full flexion can also cause nerve compression between the arcuate and medial ligaments of the floor. *3,* Ulnar nerve fibers are divided with the sensory portion more superficial; thus, pressure often causes sensory changes without motor impairment.

diskogenic disease, neurovascular compression, or pressure of the ulnar nerve at the wrist.

Radial Nerve

The radial nerve branches in the region of the elbow and is subject to entrapment. In its descent down the lateral aspect of the humerus it proceeds in front of the lateral condyle of the humerus between the brachial and the brachioradial muscles.

As the radial nerve progresses distally below the elbow joint, it passes under the origin of the short radial extensor muscle (Fig. 143). This muscle has its origin from a fibrous band that stretches from the epicondyle to the deep fascia of the volar surface of the forearm. The radial nerve divides at this point with the superficial nerve passing outside.

The deep branch passes under the fibrous band of the muscle, gives off a small recurrent branch that goes to the lateral epicondyle, and then proceeds distally to penetrate the supinator muscle through a small slit.

174

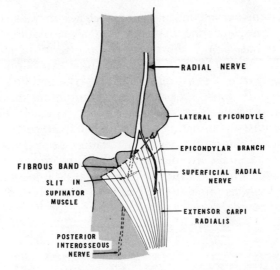

RADIAL NERVE

LATERAL EPICONDYLE

EPICONDYLAR BRANCH

FIBROUS BAND

SUPERFICIAL RADIAL
NERVE

SLIT IN
SUPINATOR
MUSCLE

EXTENSOR CARPI
RADIALIS

POSTERIOR
INTEROSSEOUS
NERVE

FIGURE 143. Course of the radial nerve. The deep nerve passes under the fibrous band origin of the short radial extensor muscle. At the division just cephalad to the fibrous band the superficial radial nerve proceeds. After entrance under the band a small recurrent branch proceeds to the lateral epicondyle.

The deep branch ultimately becomes the posterior interosseous nerve. Through its course the nerve supplies the muscles that dorsiflex the wrist and fingers.

The superficial radial nerve is exposed to direct trauma which results in sensory symptoms of pain or numbness in the lateral aspect of the forearm. Because the sensory distribution goes distally to the hand, pain may be projected to the anatomic snuffbox or the first carpometacarpal joint and mimic pathologic conditions in that area.

Fracture or dislocation of the head or neck of the radius can result in injury to the radial nerve with pain and tenderness of the entire radial nerve distribution and with motor impairment. Only the long radial extensor nerve is spared since this segment usually branches above the elbow joint.

Although direct trauma frequently is the cause of radial nerve mediated pain, usually the trauma is indirect from violent contraction of the forearm extensor muscle groups. This implies forceful or repeated motion of supination or dorsiflexion against resistance. Examples of this type of activity are playing tennis, using a screwdriver, using a heavy hammer, or any similar activity performed by a person not accustomed to doing that activity. The nerve is entrapped by the forceful contraction of the extensor group that makes taut the fibrous band at the origin of the muscles or forceful contraction of the supinator narrowing the slit

175

through which the nerve penetrates. This tension may also be referred to the lateral epicondyle via the recurrent branch and simulate tennis elbow.

Diagnosis is based on the history of the offending activity. Pain is reproduced by resisting forceful wrist and finger extension or supination. This is especially true of resisting extension of the middle finger with the elbow extended. Pressure over the site of entrapment should elicit local tenderness and characteristic radiation. Local anesthetic injection relieves tenderness and radiating pain. Because the recurrent epicondylar branch is exclusively motor there are *no* cutaneous sensory abnormalities.

TREATMENT. Usually avoidance of painful motion will suffice. This may be combined with local anesthetic and steroid injections. A splint to immobilize the wrist in neutral position has value. Persistence of symptoms may require surgery to release the epicondylar insertion of the fibrous band.

Tennis Elbow

Pain and tenderness over the lateral epicondylar region in using the forearm in motion of wrist extension and supination is commonly termed tennis elbow or lateral epicondylitis. The pain usually begins after forceful or repeated motions of wrist extension with or without supination. The condition of tennis elbow is a periosteal tear of the extensor muscles at their site of origin from the lateral epicondyle or the radiohumeral ligament (Fig. 144).

The patient notes pain or ache in the lateral elbow area after hammering, laying bricks, gardening, and so forth. The ache may subside with rest only to reappear in greater severity with repeated use of the arm until the pain and tenderness is constant and disabling.

Examination reveals full and painfree elbow motion of extension and flexion. Resisted wrist or finger extension reproduces the pain. Resisted wrist radial deviation causes pain but ulnar deviation does not. Wrist extension with fingers flexed, which eliminates the finger extensor muscles, causes pain and implicates the radial extensor muscle.

The tear may be at the muscular tendinous attachment to the lateral epicondyle, at the proximal portion of the muscle belly, or at the radiohumeral joint. Direct palpation usually can identify the site.

TREATMENT. Wrist immobilization with a cock-up splint or with a plaster cast relieves the tension on the wrist extensors. No immobilization of the elbow is indicated.

The site of tenderness can be injected with a mixture of an anesthetic agent and steroid. The exact site of pathology should be injected. The injection should *not* be into the periosteum of the lateral epicondyle, which sets up a new area of pain without improving the original pathology.

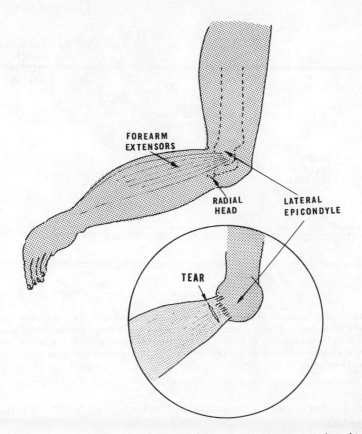

FIGURE 144. Site of tennis elbow (lateral epicondylitis). The forearm muscles which extend the wrist and fingers originate from the lateral epicondyle and extend to a ligament connecting with the head of the radius. Tennis elbow is considered to be a partial tear in the myofascial periosteal tissues of the extensor origin.

Recent literature has extolled the value of a firm elastic band 2 to 3 inches in width about the upper portion of the forearm joint below the elbow. This is worthy of trial in conjunction with other forms of treatment.

Manipulation that apparently completes or modifies the tear has been successful (Fig. 145). With the patient's elbow fully extended and the forearm fully pronated, the quick thrust is for further pronation and extension, the distance and force of the manipulation thrust being very small. The humerus must be immobilized by the therapist and the wrist must be flexed.

If the tennis elbow is in the muscle belly, local injection and splinting help, but manipulation is contraindicated.

When all conservative means fail, surgical intervention may be

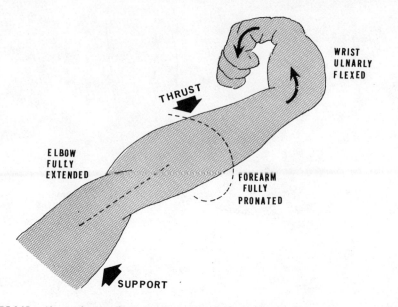

THRUST

WRIST
ULNARLY
FLEXED

ELBOW
FULLY
EXTENDED

FOREARM
FULLY
PRONATED

SUPPORT

FIGURE 145. Manipulation of tennis elbow. Procedure done with the upper arm supported, the forearm extended and fully pronated, and the wrist and fingers fully flexed in an ulnar direction. A slight brisk thrust to further extend the elbow and pronate the forearm causes a slight snap which is considered to further tear the attachment and thus give relief after subsequent healing.

requested. This either further divides the tendon or results in tenotomy or transplant.

REFERENCES

1. Cyriax, J. H.: Textbook of Orthopaedic Medicine, Vol. I: Diagnosis of Soft Lesions. Harper & Row Publishers, New York, 1954.
2. Payan, J.: Anterior transposition of the ulnar nerve: an electrophysiological study. J. Neurol. Neurosurg. Psychiatry 33:157–165, 1970.

BIBLIOGRAPHY

Basmajian, J. V.: Grant's Method of Anatomy. Williams & Wilkins, Baltimore, 1971.
Cailliet, R.: Hand Pain and Impairment, ed. 2. F. A. Davis Co., Philadelphia, 1975.
Kopell, H. P., and Thompson, W. A. L.: Peripheral Entrapment Neuropathies. Williams & Wilkins, Baltimore, 1963.
Steindler, A.: Lectures on the Interpretation of Pain in Orthopedic Practice. Charles C Thomas, Springfield, IL, 1959.
Wadsworth, T. G., and Williams, J. R.: Cubital tunnel external compression syndrome. Br. Med. J. 662–666, 1973.

CHAPTER 8

Wrist and Hand Pain

The hand is supplied by the three major nerves of the arm: median, ulnar, and radial (see Figs. 141 and 147). Because it is so richly innervated, pain can occur from either disease or trauma. The functional anatomy of the hand is discussed in *Hand Pain and Impairment*.[1]

Hilton in his treatise *Rest and Pain,* 1879, aptly stated, "The same trunk of nerves, the branches of which supply the group of muscles moving any joint, furnish also a distribution of nerves to the skin over these same muscles and their insertion and the interior of the joint receives its nerves from the same source. . . ."[2] This statement applies to all extremities but most specifically relates to the joints, muscles, and skin of the hand and wrist joint.

The dermatome mapping of the hand is depicted in Figure 146 and comprises all three nerves.

The motor innervation of the hand is supplied essentially by the median (Fig. 147) and ulnar nerves. Injury to the ulnar nerve is the most crippling in both motor and sensory aspects. This is unfortunate because the ulnar nerve is accessible to exposure and resulting trauma. Its exposure at the elbow has been described in Chapter 7.

Pain in the hand may occur from nerve compression at the wrist. Compression may involve either the median or ulnar nerve with resultant pain, anesthesia, paresthesia, weakness, and atrophy.

The radial nerve merely supplies the skin on the dorsum of the hand, the thumb, and the medial aspect of the hand to the metacarpal joints. The dermatomal region of the radial nerve is depicted in Figure 148.

CARPAL TUNNEL SYNDROME

In the carpal tunnel syndrome the median nerve is compressed within the carpal tunnel. The carpal tunnel is formed by the transverse carpal ligament which extends in two bands: from the hook of the hamate to the

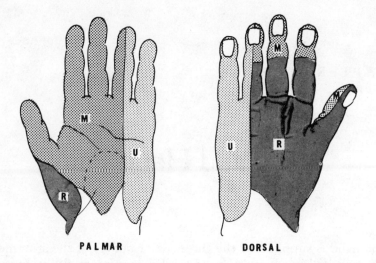

PALMAR   DORSAL

FIGURE 146.   Sensory map of the hand. Dermatomal areas of the median (M), ulnar (U), and radial (R) nerves. The dorsum of the hand is variable even to not having a radial nerve sensory area.

tubercle of the trapezius and a proximal band from the tubercle of the navicular (scaphoid) bone into the pisiform bone (Fig. 149). The floor of the tunnel is composed of the carpal bones. The contents of the tunnel are the flexor tendons of the fingers and the median nerve (Fig. 150).

Symptoms of this syndrome usually are numbness and tingling or burning of the first three fingers. The symptoms most commonly are noted in women, usually are unilateral, and characteristically occur during the night or early morning hours, awakening the patient. Relief is sought by elevating the arm, shaking the hand, or emersing the hand in hot water. Clumsiness is frequently noted in that the patient "drops things," a condition which gradually improves during the day.

Early examination reveals impaired sensation in the median nerve distribution, usually the index and middle fingers with the thumb less frequently affected. If atrophy is noted it is noted in the thenar eminence. Weakness of the thumb abductor can be elicited when the nerve compression has existed for a prolonged period of time.

The diagnosis is made by the typical history of nocturnal paresthesia, and the objective sensory findings are made by the pin prick test and by reproducing the symptoms by placing the wrist in forced flexion and maintaining this position for 60 to 90 seconds.

Confirmation of median nerve compression within the carpal tunnel may be accomplished by performing an electromyographic nerve conduction time. The nerve velocity is normal from the elbow to the wrist and then is delayed across the transverse carpal ligament into the hand and fingers.

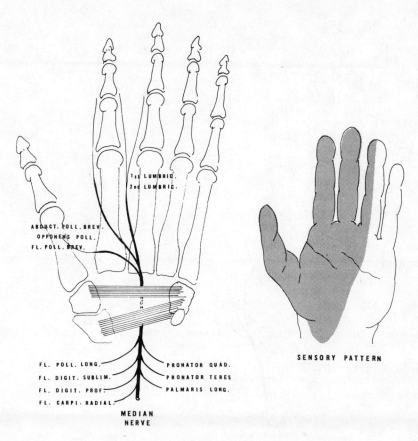

FIGURE 147. Median nerve. *Left*, Motor branches; *Right*, Sensory pattern.

The mechanism and structural changes within the nerve have recently been summarized by Sunderland.[3] The nerve is located directly beneath the retinaculum, against which it may be compressed. The walls of the tunnel are unyielding, and so compression is very possible. Within the carpal tunnel the median nerve contains approximately 24 (6 to 40) fascicles, all separated by a large amount of epineural packing. This protects the nerve from compressive forces (Fig. 151).

The perineurium has strong tensile strength and thus maintains the intrafunicular (intrafascicular) pressure. Lymphatics are contained within the epineurium; none penetrate the perineurium. Capillaries are found inside the bundles and drain into venules and veins contained within the epineurium. These are thus gradient pressure systems that must operate to maintain nutrition to the nerve axones.

The pressure within the nutrient arteries within the epineurium (P_A) must be greater than the capillary pressure within the fascicles (P_C). The

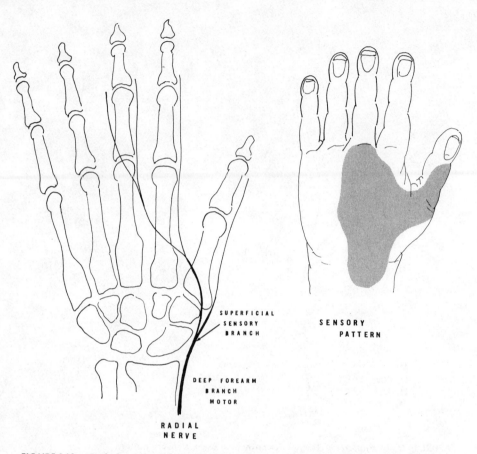

FIGURE 148. Radial nerve in the hand. The motor distribution of the nerve above the elbow supplies the triceps flexor muscle of the elbow via the brachioradial and the extensor carpi radial muscles. Below the elbow the nerve supplies ulnar wrist extensors, extensors of the fingers, and extensors of the distal phalanx of the thumb and index finger.

capillary pressure must be greater than the pressure within the fascicle (P_F) which must exceed the pressure within the veins within the epineural space (P_V) which must be greater than the pressure within the carpal tunnel (P_T). Therefore

$$P_A > P_C > P_F > P_V > P_T$$

Any change in this gradient can cause symptoms and ultimate nerve impairment.

The stages of nerve compression may be listed as

Stage I. Intrafascicular capillary distension increases intrafascicular pressure, which constricts the capillaries—a vicious cycle. Nutrition to the nerve fibers becomes impaired and the nerves become hyperexcita-

182

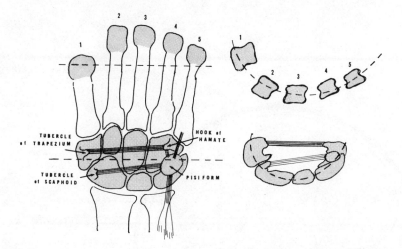

FIGURE 149. Transverse carpal ligaments. These ligaments bridge the arch of the carpal rows and form a tunnel. The proximal band extends from the tubercle of the navicular bone to the pisiform and the distal band from the tubercle of the trapezius to the hook of the hamate.

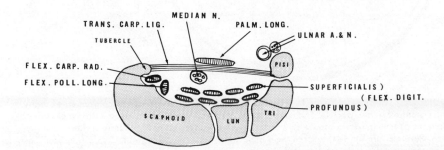

FIGURE 150. Contents of the carpal tunnel. The tunnel described in Figure 149 contains the deep and superficial long finger flexor tendons, the tendons of the long flexor muscles of the thumb and the ulnar flexor muscle of the wrist, and the median nerve.

ble. Large myelinated nerves are more susceptible than the thinly myelinated or nonmyelinated nerves explaining the paresthesia and pain. If pressure is sufficient to impair venous circulation, there is further edema and further nerve fiber impairment. This may explain the nocturnal appearance of paresthesia when the limb is hypotonic and dependent and the decrease of the paresthesia by elevating and exercising the arm. Blood pressure cuff compression aggravates the paresthesia on this basis.

Stage II. As capillary compression occurs, anoxia develops, which damages the capillary endothelium. Protein leaks into the tissues, creating more edema. Protein cannot escape the perineurium so fluid

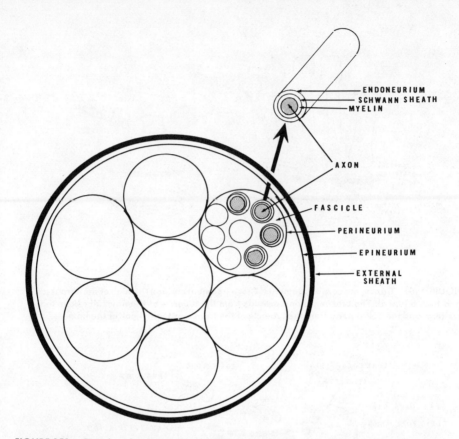

FIGURE 151. Peripheral nerve (schematic). In cross section a nerve is composed of many axons grouped into a fascicle. Each axon is surrounded by myelin enclosed within a sheath of Schwann. This is in turn coated with endoneurium which is composed of longitudinal collagen strips. Perineurium binds the fascicles which are in turn bound together by epineurium. The entire nerve is covered by an external sheath.

accumulates within the endoneurial space, which now interferes with axone nutrition and metabolism. Fibroblasts proliferate in this ischemic atmosphere and scar forms into a constrictive connective tissue. At this stage the nerve lesions are irreversible. This later stage explains the failure of long standing sensory and motor defects to improve after decompression.

In the early stages, treatment by conservative means usually affords relief. This requires splinting the wrist in a neutral position and wearing the splint *day and night*. Merely splinting the wrist during the night, because the symptoms are nocturnal, does not benefit the patient.

When there is objective evidence of motor impairment or increasing objective sensory deficit or when subjective symptoms persist, surgical intervention is indicated. Sensory return after surgical decompression is more predictable but motor return may be incomplete.

Rheumatoid arthritis with its accompanying tendinitis frequently causes nerve compression within the carpal tunnel. Surgical decompression should be considered when pharmaceutical treatment for the arthritis fails to afford relief.

Steroids injected under the carpal ligament has some diagnostic value but limited therapeutic value.

ULNAR NERVE COMPRESSION

At the wrist the ulnar nerve passes into the hand at Guyan's canal (Fig. 152), which is a shallow trough between the pisiform bone and the hook of the hamate bone. Its floor is a thin layer of ligament and muscle. The roof is the volar carpal ligament and the long palmar muscle.

After the nerve emerges from the tunnel it divides into two ulnar branches, which convey sensation from the side of the palm and the fourth or fifth fingers. The deep branch of the nerve supplies the muscles of the hypothenar eminence, the third and fourth lumbricals, all the interossei, adductor muscle of the thumb, and the deep head of the short flexor muscle of the thumb.

A lesion at the wrist may cause (1) motor and sensory deficit if the trunk is involved, (2) predominantly sensory loss if the superficial nerve is involved, or (3) primary motor deficit if the lesion is deep.

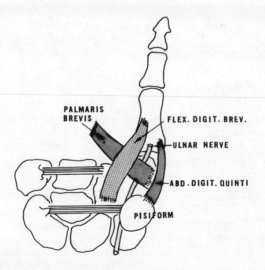

FIGURE 152. Guyan's canal. The entry of the ulnar nerve into the hand at the wrist is via a shallow trough between the pisiform bone and the hook of the hamate bone. It is covered by the volar carpal ligament and the long palmar muscle.

185

The etiologic factor usually is trauma. Symptoms usually vary from a burning sensation to uncomfortable numbness in the fourth and fifth fingers (see Fig. 146). Motor weakness is usually a clumsiness in performing fine movements with decrease of pinch strength of the thumb, adduction and abduction of the fingers, and flexion of the metacarpophalangeal joints.

Ulnar nerve symptoms must always be considered as originating at the elbow or from the cervical brachial plexus, both of which were discussed in previous chapters (Chaps. 6 and 7).

Treatment may merely require steroid and analgesic infiltration by injection into the canal. Should symptoms persist and lead to severe neurologic deficit, surgical decompression should be considered.

Pain in the wrist and hand can originate within the tendon sheaths from injury, infection, or severence. These tendons may be either flexors or extensors. The specific site of pain or the painful motion is dependent upon the specific tendon involved. Excessive repetitive movements or unphysiologic stress upon the tendon may inflame the sheath with resultant painful limited motion. The tendons usually swell and crepitation can be elicited during motion. The tendons of the wrist most commonly involved are the dorsal extensors of the wrist and the long abductor and short extensor of the thumb. This anatomic site is termed the snuffbox (Fig. 153). Tendonitis of the snuffbox is Quervain's disease.

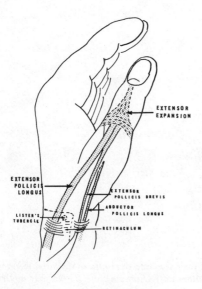

EXTENSOR EXPANSION

EXTENSOR POLLICIS LONGUS

EXTENSOR POLLICIS BREVIS

ABDUCTOR POLLICIS LONGUS

LISTER'S TUBERCLE

RETINACULUM

FIGURE 153. Snuffbox. Both the short extensor and longus abductor tendons are contained within a cuff of retinaculum. The long extensor tendon extends to the distal phalanx of the thumb in a diagonal course.

Originally described by the Swiss surgeon, Fritz de Quervain in 1895, stenosing tenosynovitis of the thumb abductor at the radiostyloid process is very common. These tendons, the long abductor and the short extensor, move in a common sheath that passes in a bony groove over the radiostyloid process (Fig. 154). Distal to the process the tendons form a sharp angle of approximately 105 degrees that encourages friction with resultant synovitis. During thumb pinching activities the long abductor muscle stabilizes the thumb, which causes friction.

Symptoms are of an aching discomfort over the styloid process aggravated by movements of the wrist and thumb. Abduction of the thumb against resistance can reproduce the symptoms and there may be

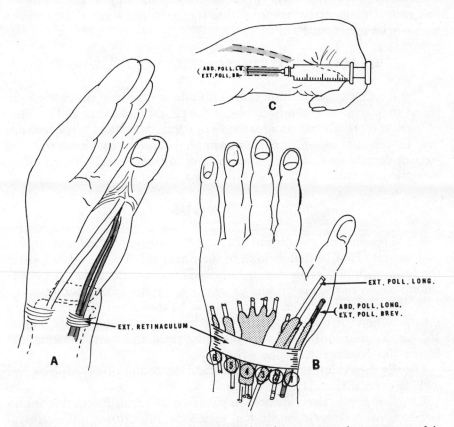

FIGURE 154. Quervain's disease. A, The combined tendons pass over the prominence of the radiostyloid process; B, The extensor tendons of the fingers pass under the retinaculum; C, Method of steroid injection.

Sites of tendon injection in B are 1, long thumb extensor; 2, radial wrist extensor; 3, long thumb abductor; 4, finger extensors; 5, little finger extensor; and 6, ulnar wrist extensor.

187

tenderness over the tendon. A diagnostic procedure is to reproduce the symptoms by flexing the thumb and cupping it under the fingers and then flexing the wrist in an ulnar direction. This stretches the thumb tendons and causes pain.

The pathology is a combination of edema within the sheath and increased vascularity of the sheath. The sheath constricts the tendons and restricts movement. Fine adhesions have been described when the synovitis persists.

Treatment demands immobilization of the thumb and wrist in a padded mold cast. Injection of steroids into the sheath are beneficial (Fig. 154, B and C).

If after three to four weeks of immobilization and a series of injections there is persistence of symptoms, surgical decompression is indicated. It must be ascertained at surgery that both tendons (long thumb abductor and short thumb extensor) are within the sheath or surgical decompression will be ineffectual.

TEAR OF THE LONG EXTENSOR TENDON OF THE THUMB

The long extensor tendon of the thumb can rupture because of the sharp angulation of the tendon about the tubercle of Lister and because of friction there. Rheumatoid arthritis or damages from a Colles' fracture may hasten rupture. This rupture can be diagnosed by virtue that the patient cannot extend the distal phalanx of the thumb. Conservative treatment for this is ineffectual and surgical repair is indicated.

TRIGGER FINGERS

A sudden snapping sensation of a finger during flexion and reextension may be noted and actual locking of the finger may result. Once locked, the finger cannot further flex or extend. This condition can occur in the thumb or any of the fingers and usually ocurs in the flexor tendons.

A nodule forms on the tendon within its thickened synovium-lined sheath. When the nodule becomes too thick, obstruction occurs. Figure 155 depicts the more common snapping third and fourth fingers and Figure 156 depicts the snapping thumb.

Usually the nodules can be palpated and the condition of snapping and locking demonstrated by the patient.

Cortisone injection within the sheath may result in complete and permanent recovery. Should a series of injections fail, surgical intervention relieves the problem. The annular band (see Fig. 155) is slit to permit the nodule to pass, whereas excision of the nodules may cause a larger nodule to reform.

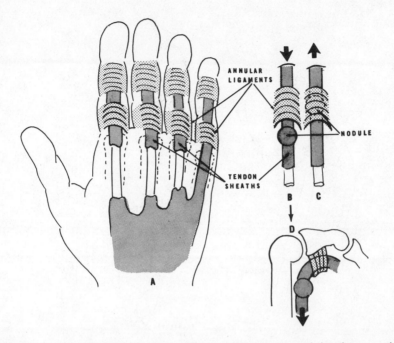

FIGURE 155. Trigger finger. *A*, The flexor tendons within their synovial sheath pass under the annular ligaments at the metacarpal heads; *B*, Nodule proximal to the ligament prevents flexion; *C*, Nodule is trapped; *D*, Re-extension is prevented.

SPRAINS AND DISLOCATIONS

Sprains may be momentary subluxation which reduces spontaneously. As they reduce, x-ray pictures tend to be normal and the soft tissue injury escapes detection. The capsule and collateral ligaments may be torn and actual articular dislocation may occur.

Joint limitation, pain, and swelling may be evident on examination. Passive range of motion when compared to the normal contralateral joint reveal excessive mobility, but usually joint restriction is evident.

When a sprain is suspected, immobilization in a slight flexed position of function for two to three weeks will usually permit the soft tissue injury to heal (Fig. 157). The immobilization should be followed by progressive active exercise.

Subluxation of the metacarpophalangeal joint may tear the palmar plate (Fig. 158). Palmar plates are fibrocartilaginous plates that reinforce the joint capsule to prevent hyperextension and prevent excessive friction of the flexor tendons. The distal portion of the plate is cartilaginous and the proximal portion is membranous. The distal portion is firmly

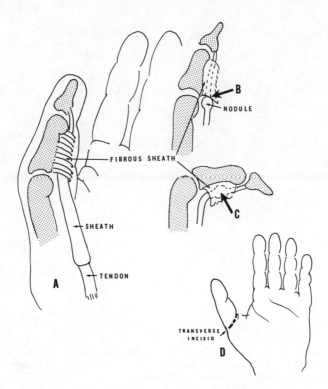

FIGURE 156. Snapping thumb flexor. *A*, Flexor tendon under its sheath passing through the fibrous canal; *B*, Nodule which prevents flexion; *C*, Nodule, trapped within sheath, which prevents re-extension; *D*, Site and direction of decompressing the tenosynovitis.

attached to the phalanx, but the proximal portion is loosely attached to the metacarpal bone.

The palmar plate, because of its membranous portion, will retract when permitted to remain in a shortened position and thus form contracture. Prolonged immobilization of a finger in flexion results in a fixed flexion contracture.

RHEUMATOID ARTHRITIS

Arthritis of the numerous joints of the hand is a frequent cause of pain. Arthritis may be inflammatory, infectious, or degenerative.

Rheumatoid arthritis is initially a disease of soft tissue, a disease of synovium. Patients with rheumatoid arthritis have a high incidence of involvement of tendons and their sheaths. As the disease tenosynovitis proceeds, granulomatous synovium invades the tendon, thus causing it to weaken, lengthen, and frequently rupture. Simultaneously the synovitis invades all the periarticular tissues, the capsule ligaments, and ultimately

190

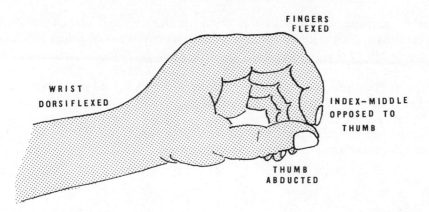

FINGERS
FLEXED

WRIST
DORSIFLEXED

INDEX—MIDDLE
OPPOSED TO
THUMB

THUMB
ABDUCTED

FIGURE 157.   Functional position of the hand. This is the ideal position to be attempted in splinting. Specific medical indications may require some deviation.

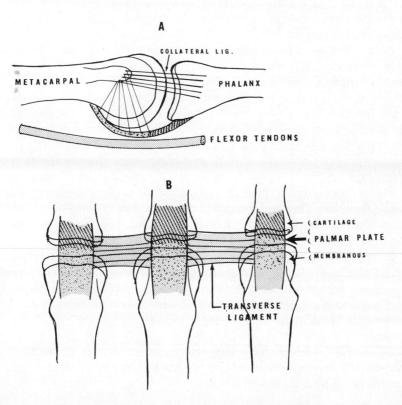

A

COLLATERAL LIG.

METACARPAL

PHALANX

FLEXOR TENDONS

B

CARTILAGE
PALMAR PLATE
MEMBRANOUS

TRANSVERSE
LIGAMENT

FIGURE 158.   Palmar plate. There is a fibrocartilaginous plate on the palmar surface of the joints that reinforces the capsule. A, The proximal portion is membranous and is loosely attached to the metacarpal. The distal portion is cartilaginous and is firmly attached to the phalanx. B, The plates are connected to the deep transverse ligament. This ligament also prevents lateral motion of all fingers but the thumb.

191

the cartilage. Full discussion of the pathology and manifestation of rheumatoid arthritis in regards to each specific joint or joints is beyond the scope of this book.

The joints of the hand most frequently involved are the proximal interphalangeal and metacarpophalangeal joints as well as the wrist joint. Loss of the proximal interphalangeal joint and the metacarpophalangeal joints are far more disabling than loss of the distal phalangeal joints. Impaired thumb motion causes a major functional handicap. If the carpometacarpal and metacarpophalangeal joints of the thumb are involved, rotation of the thumb is restricted and tip to tip opposition is lost.

Treatment

Treatment of an acute condition of the rheumatoid hand is directed towards the inflammation, swelling, and pain. Obviously the major effect of treatment is towards the systemic disease, but the hand must be considered seriously.

Rest of the hand is mandatory but is extremely difficult to accomplish. Movement during the acute phase is potentially structurally detrimental, yet activities of daily living require some use of the hands. Splints to mobilize the hand and fingers are cumbersome, difficult to apply, and maintain and frequently tax the patient's tolerance and cooperation. Splints can be applied for varying periods during the day and for night wear, but the deforming forces continue during the unsplinted periods. See my earlier book, *Hand Pain and Impairment,* for details of specific splinting.[1]

Heat is soothing and permits the patient to perform relatively painless active exercise. This maintains full range of motion and delays atrophy from disease and disuse. Ice, rather than heat, has many advocates and can be used if better tolerated and accepted by the patient.

Joint protection activities must be observed.[4] These are activities that avoid the deforming movements of the hand and wrist, such as squeezing which overuses the flexors or jar turning which places hand and fingers in an ulnar deviation position. Forced flexion can cause volar subluxation of the metacarpophalangeal joint and must be avoided.

Isometric exercises that encourage muscle contraction without joint motion are desirable and can be taught to a member of the family or administered by a therapist.

BOUTONNIERE AND SWAN NECK DEFORMITIES

Boutonniere and swan neck deformities occur as a result of soft tissue defects and abnormal muscular action upon the joints of the fingers.

Splinting, exercises, and surgical intervention are currently relatively ineffective, but total joint replacements may ultimately be more effective.

DEGENERATIVE JOINT DISEASE

The most common and most disabling joint arthritis in relatively young people is degenerative disease of the thumb carpometacarpal joint (Fig. 159). In this condition there is tenderness, stiffness, pain, and crepitation on movement. Grip becomes impaired and fine movements requiring

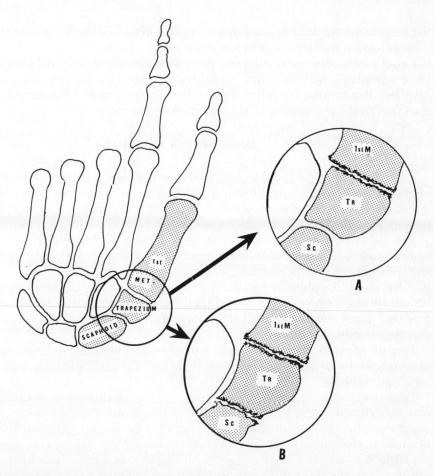

FIGURE 159. Degenerative joint disease at the base of the thumb. A, Degenerative arthritis of the first carpometacarpal joint responds well to treatment by splinting, injection, or surgery (excision of trapezium fusion or implant). B, When there are arthritic changes in the trapezioscaphoid joint, arthrodesis or implant may not be of value.

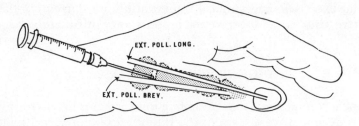

FIGURE 160. Technique of injection of metacarpotrapezium joint. Flexing the thumb opens the joint, which can be palpated. The needle is inserted within the confines of the snuffbox.

thumb tip-to-tip activity are restricted. This condition usually is bilateral and is more prevalent among women.

Relief is afforded by resting the part. This can often be done with a leather or plastic molded splint. Intra-articular injections of steroids give excellent but temporary relief (Fig. 160). Surgery ranging from fusion, resection of the trapezium, or silicone implant has value.

VASCULAR IMPAIRMENT OF THE HAND

Raynaud's phenomenon is a condition in which there is spasm of small blood vessels, especially of the fingers. The patient may observe intermittent attacks of sudden pallor to actual blanching of the fingers. Pain may result. The acute pallor episode may be followed by cyanosis.

The condition is attributed to arterial spasm, a manifestation of vasomotor instability. It is usually triggered by cold or emotional stress and is most prevalent in middle-aged females.

A Raynaud-like phenomenon of the hands can result from neurovascular compression of the cervical dorsal outlet, may be the result of repeated occupational trauma such as operation of a pneumatic drill, or may be a familial condition.

Treatment is prophylactic. Smoking must be eliminated, handling iced objects must be avoided, and warm clothing and gloves should be worn in inclement weather.

Painful vascular tumors may exist that may escape detection, yet they may be exquisitely painful and disabling. The glomus tumor is such an example. This is a subungual tumor that may be tender, painful, and sensitive to temperature changes. At first the tumor may not be visible or palpable but, when it finally appears, it does so as a blue spot in the subungual region or a ridge in the nail. The only treatment for the glomus tumor is total surgical excision.

REFERENCES

1. Cailliet, R.: Hand Pain and Impairment, ed. 2 F. A. Davis Co., Philadelphia, 1975.
2. Hilton, J.: Rest and Pain. Wm. Wood & Co., New York, 1879.
3. Sunderland, S.: The nerve lesion on the carpal tunnel syndrome. J. Neurol. Neurosurg. Psychiatry 39:615–626, 1976.
4. Melvin, J. L.: Joint Protection training and energy conservation. In Rheumatic Disease: Occupational Therapy and Rehabilitation. F. A. Davis Co., Philadelphia, 1977.

BIBLIOGRAPHY

Allen, E. V., and Brown, G. E.: Raynaud's disease. A clinical study of 147 cases. J.A.M.A. 99:1472, 1932.

Boyes, J. H. (ed.): Bunnell's Surgery of the Hand, ed. 5. J. B. Lippincott, Philadelphia, 1970.

Chase, R. A.: Surgery of the hand. N. Engl. J. Med. 287:1174, 1972.

Carroll, R. E., and Berman, A. T.: Glomus tumors of the hand. J. Bone Joint Surg. 54[Am.]:697, 1972.

Carstam, N., Eiken, O., and Andren, L.: Osteoarthritis of the trapezio-scaphoid joint. Acta Orthop. Scand. 39:354, 1968.

Flatt, A. E.: The Care of the Rheumatoid Hand. C. V. Mosby Co., St. Louis, 1963.

Kapell, H. P., and Thompson, W. A. L.: Peripheral Entrapment Neuropathies. Williams & Wilkins, Baltimore, 1963.

Wynn Parry, C. B.: Rehabilitation of the Hand. Butterworths, London, 1966.

Zancolli, E.: Structural and Dynamic Basis of Hand Surgery. J. B. Lippincott, Philadelphia, 1968.

CHAPTER 9

Hip Joint Pain

The hip joint in man is predominantly weight bearing and intrinsically involved in ambulation. It is well constructed and, in spite of the numerous and varied trauma imposed on it, it only infrequently becomes impaired.

Pain and disability from hip involvement can benefit from proper evaluation and medical and physical management. Surgical management has made great strides recently in alleviating pain and improving function, but many nonsurgical therapeutic approaches may eventually eliminate need for surgical intervention or make ultimate surgical procedures more effective.

ANATOMY

The head of the femur is spherical and points medially upward and forward (Fig. 161). It articulates with the acetabulum but, because of the angle of its head and neck, the anterior portion of the head of the femur is not engaged in the socket in the neutral leg position.

The acetabulum is horseshoe shaped and is covered peripherally with cartilage. The center of the horseshoe is not covered with cartilage. The open lower portion of the acetabulum is completed into a ring by the transverse acetabular ligament (Fig. 162, above). The acetabulum is deepened by a complete ring of fibrocartilage termed the labrum. The head of the femur is held firmly in the acetabulum by a thick capsule (Fig. 162, center). The fibers of the capsule are oblique and become taut when the hip is extended. There are portions of the capsule that thicken to form ligaments. Termed for their site of origin they include the iliofemoral, pubofemoral, and ischiofemoral ligaments (Fig. 162, below).

In erect stance the center of gravity passes behind the center of the hip joint. The pelvis must therefore rotate backwards. To resist this rotation

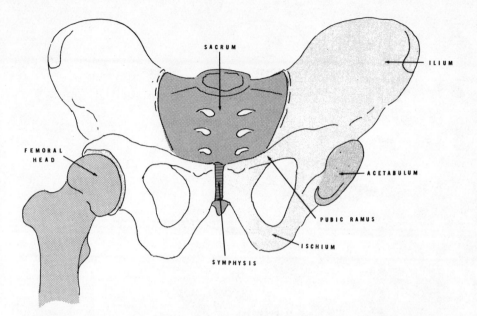

FIGURE 161. Anterior view of bony pelvis. The pelvis is pictured anteriorly with the left femur omitted to reveal the acetabulum.

the anterior portion of the capsule is thickened to form the iliofemoral ligament.

In standing with toe-out stance the head of the femur is directed forward out of the socket. The iliofemoral ligament moves laterally and exposes an anterior portion of the hip joint to no ligamentous support. This exposed area is covered by the tendon of the psoas muscle. The portion of the joint that does not have cartilage is lined with synovial membrane, which extends to completely encircle the neck of the femur.

There is a ligament that connects the head of the femur to the center of the acetabulum within the centrum of the hip joint. This ligament is a hollow tube of synovial membrane that transmits blood vessels to the head of the femur: the branches of the medial circumflex and obturator arteries.

The blood supply to the hip joint is from the medial femoral circumflex, lateral femoral circumflex, and obturator arteries and branches from the gluteal artery.

HIP JOINT MOTIONS

The hip joint motions are depicted in Figures 163 and 164. Flexion is limited by the hamstring muscles when the knee is extended and by

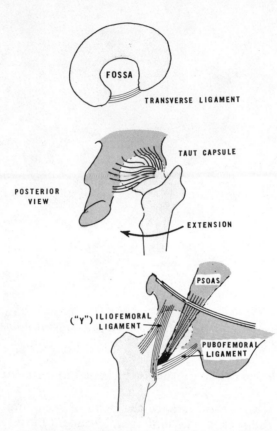

FIGURE 162. Hip joint. *Above,* The full articular circle of the acetabulum is completed in the inferior aspect by the transverse acetabular ligament. *Center,* The oblique capsular fibers become taut as the hip extends. *Below,* The anterior capsule is reinforced by the iliofemoral, pubofemoral, and ischiofemoral ligaments and the psoas tendon.

abdominal wall contact with the knee flexed. Extension is limited by the ligamentous thickening of the fibrous capsule. With the hip extended the capsular fibers limit internal and external rotation. Abduction is limited by the adductor muscle group and adduction by the tensor muscle of the fascia lata and the abductor muscle group.

During relaxed standing the hip is fully extended. If the hip is hyperextended this position is achieved by rotation of the pelvis and extension of the lumbar spine into further lordosis.

Walking

Walking exerts repeated stretching of the hip capsule, ligaments, fascia, and muscles of the flexor aspect of the hip joint. This is so because every

198

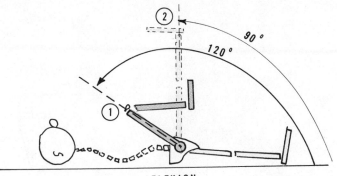

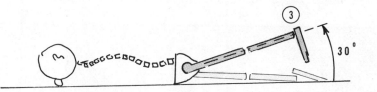

FLEXION

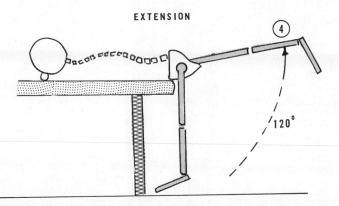

EXTENSION

FIGURE 163. Hip joint range of motion. 1, Hip flexion with knee flexed from 0 to 120 degrees limited by contact of the thigh with the abdominal wall; 2, Hip flexion with knee extended 90 degrees is limited by the hamstring muscles; 3, Hip extension 30 degrees with patient prone; 4, With opposite leg flexed to 90 degrees by patient lying prone on examination table, hip flexion of tested leg should go 90 to 120 degrees. Hip flexion contracture can be measured clinically by this position.

alternate step during gait requires that the leg be fully extended (Fig. 165). The natural tendency of fibrous connective tissue to shorten causes hip flexion to become contracted and resist full extension. Habitual sitting and lack of exercising predisposes to hip flexor contracture with

199

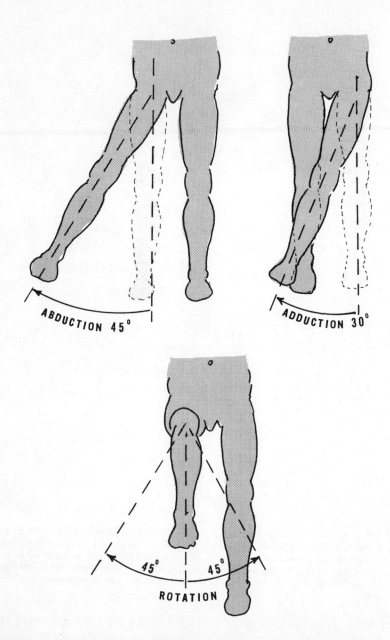

FIGURE 164. Hip joint range of motion. Hip abduction is normally 45 degrees from 0 degrees full extended leg and adduction is 30 degrees. Rotation of hip measured with knee flexed is normally 45 degrees of internal rotation and 45 degrees of external rotation.

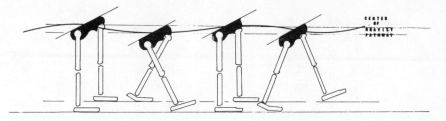

FIGURE 165. Undulant course of pelvic center in gait without determinants. The marked vertical undulations of the center of gravity occur when gait is performed with stiff knees and no lateral motion of the pelvis.

resultant impairment of gait and adverse effect upon the lumbar spine. As flexion contracture occurs the iliotibial band similarly contracts, thus producing abduction and external rotation of the hip as well as flexion contracture.

In normal walking the determinants of gait that level out the center of gravity translation are pelvic rotation (Fig. 166), pelvic tilt (Fig. 167), lateral displacement of the pelvis, and knee flexion in the stance phase (Fig. 168). All these determinants require adequate hip range of motion.

In normal gait the femur rotates upon the pelvis and the tibia upon the femur. This also requires normal hip range (Fig. 169). Loss of hip motion impairs gait and decreases efficiency, grace, and conservation of energy.

The average normal gait reports 60 percent of its timing in the stance phase and 40 percent in the swing phase (Fig. 170). Patients with hip disease spend a disproportionate amount of time in the stance phase of the involved extremity. Patients with diseased hips also excessively rotate laterally over the affected hip during stance. These variations change the gait velocity, stride length, and cadence in the individual's attempt to minimize pain and improve stability. What these changes add to the pathology of the head of the femur or the acetabulum is not determined.

Standing

Weight upon the hip joint in standing has been calculated. Using a lever system, Figure 171 indicates the distance from the body weight center of gravity is approximately 4 inches to the center of fulcrum at the hip joint. The abductor gluteal muscles must exert a balancing force about the fulcrum through a lever arm of 2 inches. It is thus apparent that a 150 pound individual imposes a pressure of 450 pounds upon the femoral head. This force, which is normally sustained by the femoral head, indicates why such tremendous forces are required to cause fracture and why the engineering studies of artificial hips have been so difficult. In diseased hips added body weight of the individual must also be carefully controlled as each pound of weight gain imposes three pounds upon the

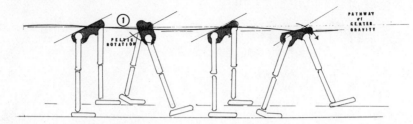

FIGURE 166. Pelvic rotation: determinant of gait. The pelvis rotates forward with the swinging leg aimed to decrease the angle of the leg to the floor measured at the hip joint. This decreases the vertical undulation of the center of gravity.

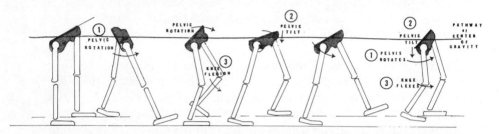

FIGURE 167. Pelvic tilt: determinant of gait. As the left leg swings through, the pelvis drops on the left (2). The left hip and knee flex (3). The last figure depicts the right leg swinging through with the right pelvis dipping and the right hip and knee flexing. This decreases the center of gravity undulation, but requires full hip range of motion.

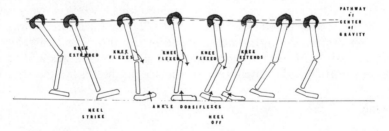

FIGURE 168. Knee flexion during stance phase: determinant of gait. The knee is fully extended at heel strike. As the body passes over the center of gravity the knee flexes to decrease the vertical amplitude of the pathway of the center of gravity. The knee reextends at the end of the stance phase: the "heel off."

hip. A weight gain of 50 pounds would impose 150 pounds upon the hip.

Use of a crutch or cane in the opposite hand markedly decreases the pressure upon the hip as described in Figure 172. Assuming the patient holds the cane 20 inches from the center of gravity and presses with 30

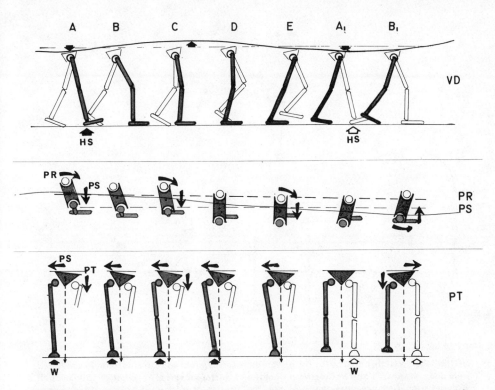

FIGURE 169. Composite schematic determinants of gait. *VD* indicates the vertical displacement of the pelvis from the side view. *PR* is pelvic rotation viewed from above as the left leg swings through. *PT* depicts pelvic tilting. The bottom figure shows the weight bearing leg *(W)* going into a Trendelenberg position as the hip adducts. *PS* shows the pelvic shift. All these determinants require good hip motion.

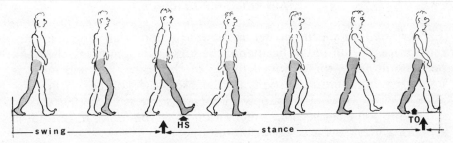

FIGURE 170. Gait. The shaded leg depicts the swing phase ending at heel strike *(HS)* and the stance phase continuing until toe off *(TO)*. The hip extends at the beginning of swing and continues through midstance phase.

203

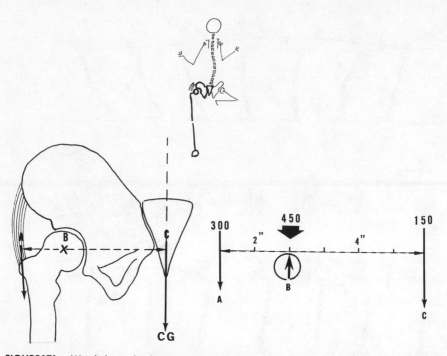

FIGURE 171. Weight borne by the hip on standing. *Left,* Weight upon the hip joint(B) combined from body weight (C) and balancing pull of the abductor muscles (A). *Right,* A 150 lb. adult with a 4 inch distance from center of gravity to hip joint fulcrum is balanced by gluteal muscular action 2 inches from the fulcrum. The glutei exert 300 lb. to balance, thus total pressure upon the hip joint is 450 lb.

pounds of pressure, the pressure upon the hip joint (of a 150 pound person) is only 120 pounds because there is less need for gluteal muscular balance. The cane acts in a clockwise direction as do the gluteals, thus they balance each other.

PAIN IN THE HIP JOINT

There are four specific structures about the joint that can elicit pain: the fibrous capsule and its ligaments, the surrounding muscles, the bony periosteum, and the synovial lining of the joint. The cartilage is insensitive, but there is question as to the insensitivity of the subchondral bone by virtue of its blood supply with its sensory vasomotor nerves.

The hip joint has sensory innervation from the femoral, obturator, superior gluteal, and accessory obturator nerves.

The nerve branches of the femoral nerve arise directly from the nerve and supply the iliofemoral ligament near its femoral attachment. The articular branch of the obturator supplies the medial portion of the capsule. The superior gluteal nerve gives branches to the fibrous layer of

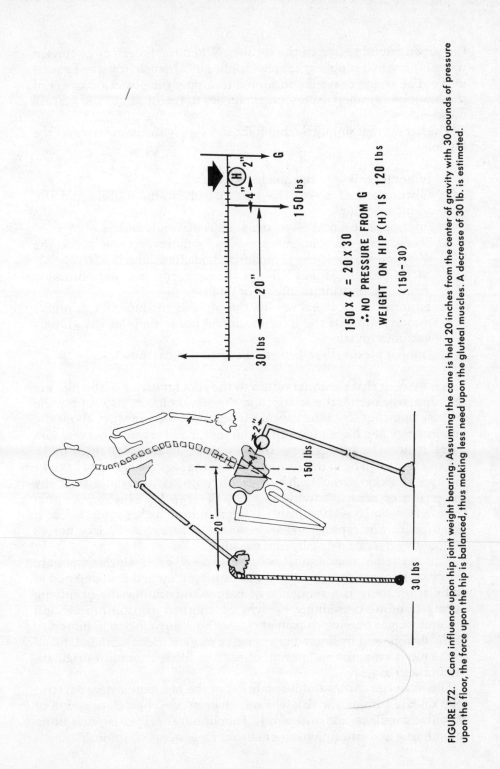

FIGURE 172. Cane influence upon hip joint weight bearing. Assuming the cane is held 20 inches from the center of gravity with 30 pounds of pressure upon the floor, the force upon the hip is balanced, thus making less need upon the gluteal muscles. A decrease of 30 lb. is estimated.

$150 \times 4 = 20 \times 30$

∴ NO PRESSURE FROM G

WEIGHT ON HIP (H) IS 120 lbs

(150 − 30)

30 lbs

20″

150 lbs

4″

2″

G

H

205

the superolateral region of the capsule. When the accessory obturator nerve is present it supplies the area supplied by a branch from the femoral nerve. The sciatic nerve is considered to supply the posterior aspect of the hip joint capsule via muscular branches to the quadrate and gemelli muscles.

The nerves that supply the hip joint also supply the muscles about the hip joint.

1. Femoral nerve—to the quadriceps muscle
2. Obturator nerve—to the external obturator muscle and the adductor group
3. Inferior gluteal nerve—to the gluteus maximus muscle
4. Sciatic nerve—semimembranous and semitendinous as well as the great adductor and the gemellus and quadrate muscle of thigh
5. Sacral plexus (S_1 and S_2)—to the piriform, internal obturator, gemelli, and quadrate muscle of thigh
6. Superior gluteal nerve—to the gluteus medius and minimus muscle, whereas the inferior gluteal nerve supplies the gluteus maximus muscle
7. Lumbar plexus (L_2 to L_4)—to the greater psoas muscle

It is evident that the innervation to the short muscles of the hip, i.e., the obturator nerve, the sciatic, and the sacral plexus, also supply the sensory branches from the capsule. The cutaneous branches about the hip originate at a higher level than the motor and capsular nerves. The lateral femoral cutaneous nerve that covers the anterolateral thigh are L_2, the anterior thigh from the continuation of the femoral nerve by L_2 to L_4, the upper portion of the thigh by the iliohypogastric, and the buttocks by the posterior primary division of D_{12} (T_{12}) to L_3. This implies that superficial cutaneous abnormality is referred from higher spinal levels. It also explains why capsular irritation via the obturator and sciatic nerves frequently refers pain distally into the knee.

Pain in the hip region must be differentiated as to which tissues are involved. Because the hip is a weight-bearing joint and instrumental in ambulation there is a sequence of pain with pain initially occurring during standing or walking (weight bearing) to pain on hip motion without weight bearing to pain at rest without any motion. Pain is also gradually followed by limited hip range of motion. Because the cartilage of the hip is avascular and devoid of sensory fibers pain must originate from other tissues.

The most common painful condition of the hip joint is degenerative joint disease (osteoarthritis), which is characterized by deterioration of articular cartilage and ultimately subchondral sclerosis with bone remodeling and osteophyte formation. Degenerative joint disease is considered primary when it is the result of aging alone but is secondary when trauma or systemic factors are involved. Many factors may be involved: genetic predisposition, alteration of lubrication, abnormality of cartilage, and metabolic diseases.

The articular cartilage is composed of four layers. The superficial layer consists of flattened chondrocytes surrounded by tightly woven bundles of collagen fibers lying parallel to the subchondral bone. The intermediate zone has random arrangement of collagen fiber intertwining around chondrocytes. The deeper midlayer has the collagen fibers perpendicular to the surface. This is the layer that permits compression and resiliency of the tissue. The deepest layer is the transition layer between cartilage and bone. It is calcified with minimal fibers and cells.

The cartilage is avascular and depends upon imbibition and diffusion for its nutrition. Much of the nutrition is afforded by the synovial fluid. Diffusion of nutritive fluids through the endochondral plates via the bone blood vessels is accepted but not yet fully understood.

Lubrication of cartilage upon cartilage is considered to be from a glycoprotein faction that is pumped out of cartilage by pressure. Motion and intermittent compression of cartilage is thus mandatory for adequate nutrition. Cartilage degenerates because of subchondral sclerosis which impairs diffusion. There is, thus, a decrease in the lubricant, which enhances superficial damage.

It has been postulated that repeated forceful perpendicular impact on the limb causes microfractures of the cancellous endochondral bone which, as they heal, occlude the permeability of nutritive fluids to the cartilage and thus predispose to osteoarthritis.

The cartilage itself does not have a nerve supply. The sensory nerve endings are found in the capsule and the ligaments but none penetrate the intact cartilage. The synovial membrane is well innervated, but most nerves innervate the blood vessels and very few are sensory for pain transmission.

The synovial membrane is probably involved very early in arthralgia with hyperemia. Ultimately the membrane increases in both quantity and composition of synovial fluid. This synovial inflammation affects the autonomic nerves, which are with sensory and vasomotor nerves, so that pain and further congestion results. Synovium has been shown to be sensitive to hyperemia and distension.

The capsule also plays a vital part in the production of pain. The capsule becomes infiltrated by inflammatory cells with gradual thickening.

Usually the capsule is tense, but it is relaxed in flexion, abduction, and external rotation. The iliofemoral and the ischiofemoral ligaments become taut in hip extension, adduction, and internal rotation. Since the external rotator (the gemelli and quadrate muscles) innervation is closely related to the innervation of the capsule it becomes of clinical significance that, in hip pathology involving the capsule, flexion remains unrestricted longer than does rotation and hip extension. As the joint surfaces change in contour and depth, abduction and external rotation become increasingly more difficult with progressive limitation and proportional pain.

Circulatory impairment is also considered to cause hip pain. These vascular changes occur not only in the articular and periarticular tissues but also in the bone and subchondral tissues. The complete relationship of vascular impairment to hip degeneration and pain is not currently clear.

Management of Hip Pain

The usual effective measures to decrease hip pain can be divided into nonsurgical and surgical.

NONSURGICAL MEASURES. The nonsurgical measures are rest, immobilization by casting, traction, intra-articular injection of steroid or anesthetic agent, and chemical denervation.

Rest helps minimize or eliminates weight bearing. This can vary from restricted standing or walking to elimination of jumping or running to complete bedrest. Use of a cane(s) in the opposite hand or crutch(es) also decreases weight bearing. Loss of excessive body weight is always indicated.

Proper positioning during prolonged bedrest must be assured. Full range of motion must be maintained during bedrest to prevent contracture weakness or atrophy. Flexion contracture is by far the most common problem; therefore it should be prevented.

Lying in the prone position frequently during the day helps stretch the hip flexors. A pillow under the thigh with the patient in the prone position is of value providing it does not cause excessive lumbar lordosis. Hip extensor exercise to strengthen the extensor musculature is of value (Figs. 173, 174, and 175).

Immobilization by casting will rest the inflamed joint. A spica-type cast usually is effective and must include the pelvis. Keeping the knee free and exercising the quadriceps during cast application is indicated.

Traction is effective in immobilizing the hip joint and elongating the capsule of the hip, thus separating the articular surfaces. This traction can be applied by pin insertion or by skin application (Fig. 176).

Intra-articular injection of a steroid or anesthetic agent requires special skill. Usually the joint can be entered by measuring 2 to 3 cm. below the anterosuperior iliac spine and entering 2 to 3 cm. lateral to the femoral artery (identified by pulsation) (Fig. 177). The needle proceeds

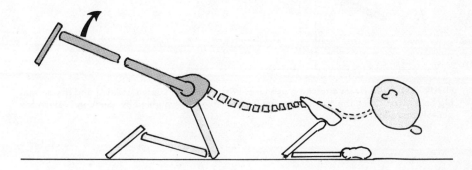

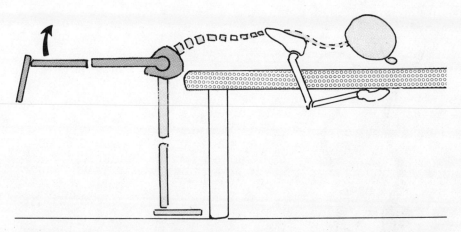

FIGURE 173. Exercises to extend hip joint: exercises aimed to stretch the anterior hip capsule and strengthen the extensor musculature. *Above,* With patient prone, the leg is extended preferably against resistance. A pillow under the abdomen decreases excessive lordosis. *Center,* With contralateral knee flexed and bearing weight the involved leg is extended. *Below,* Similar to the center illustration except the patient is prone over table. The dependent leg stabilizes the pelvis and the lumbar lordosis is decreased.

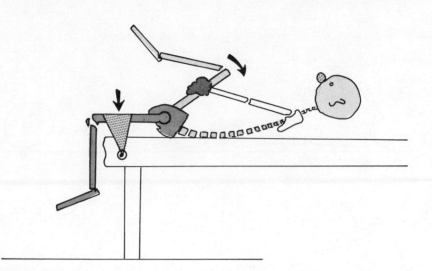

FIGURE 174.   Hip flexor stretching exercise. With patient in supine position and the normal hip held to the chest, the opposite leg by its own weight or weighted by sandbag is extended actively and passively.

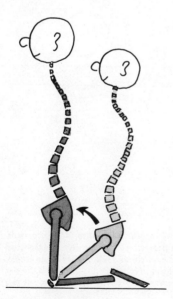

FIGURE 175.   Hip extensor exercise. From the full kneeling position the patient arises to full erect kneeling posture. This stretches the hip flexors and strengthens extensors.

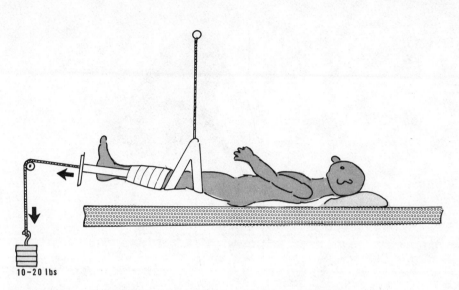

FIGURE 176.   Technique of hip traction by skin application to lower leg.

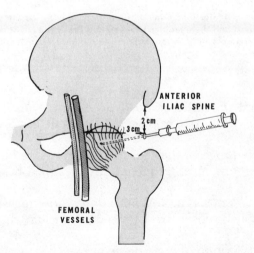

**ANTERIOR
ILIAC SPINE**

2 cm

3 cm

**FEMORAL
VESSELS**

FIGURE 177.   Technique and site of intra-articular injection. The needle is inserted 2 to 3 cm. below the anterosuperior iliac spine and 2 to 3 cm. lateral to the femoral artery pulsation and penetrates in a posterior medial direction (a 60 degree angle) until bone is reached. After aspiration, which may aspirate fluid, the steroid-anesthetic agent is injected.

211

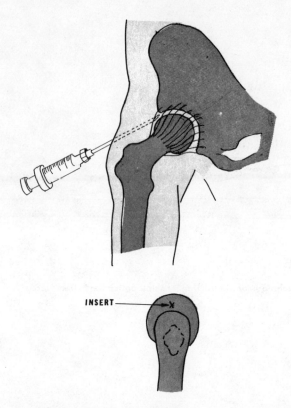

FIGURE 178. Intra-articular injection. While patient is lying on the contralateral hip the trochanter is palpated and the needle is inserted midline of the superior aspect of the femur neck—to follow along neck until the capsule of the hip joint is reached. Aspiration of fluid is followed by a steroid-anesthetic agent.

posteriorly and medially at 60 degrees angle until the capsule is penetrated (Fig. 178). Aspiration should precede injection. Fluoroscopy may be helpful.

Chemical denervation of the capsule requires nerve blocks by using an anesthetic agent and most frequently involves the obturator nerve (Fig. 179).

SURGICAL INTERVENTION. When conservative measures fail, surgical intervention may be considered. Procedures include the following:

1. Revascularization of bone procedures.
2. Osteotomies to alter alignment of head to acetabulum.
3. Denervation.
4. Arthrodesis.
5. Arthroplasties of cup or total hip replacement.

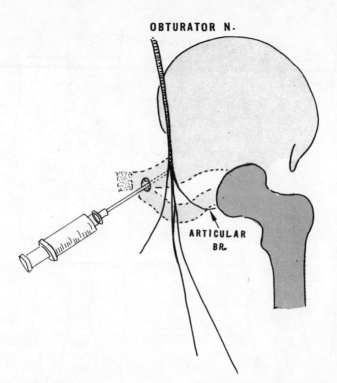

OBTURATOR N.

ARTICULAR BR.

FIGURE 179. Technique and site of obturator nerve block. With patient supine and thighs separated, a wheal is raised 1 cm. lateral to the pubic tubercle. A 22-gauge 8 cm. needle is directed perpendicularly until the inferior ramus of the pubis is reached. The direction of the needle is then changed into a lateral and superior direction, parallel with the pubic ramus, several centimeters. After aspiration 5 to 10 cc. anesthetic agent is injected. An effective injection is determined by adduction and external rotation paresis and *not* by an area of anesthesia.

PAIN FROM OTHER SITES

Pain claimed to be in the hip area by the patient may be referred there from other sites and must always be included in the differential diagnosis.

Myofascial pain from distant trigger areas within muscle, myofascial tissues, the fascia lata, or ligaments may occur (Fig. 180). These trigger areas may be local or distant and respond well to local injection using an anesthetic agent, deep massage, or even vasocoolant spray.

Pain may be referred to the hip (buttocks) area from the lumbar spine. In this condition the pain can be reproduced by lumbar movement. Usually hyperextension of the lumbar spine reproduces the pain more than does flexion. Pain and tenderness are usually elicited by pressure upon the sciatic nerve in the vicinity of the sciatic notch. Straight leg

213

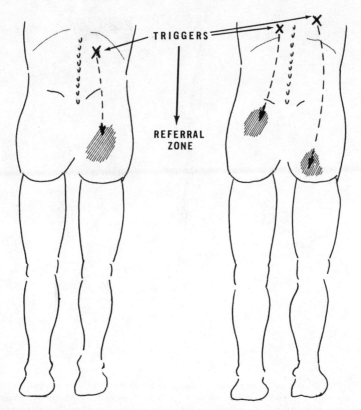

FIGURE 180. Pain referred to hip region. Radiating pain may occur from irritation of the nerves emanating from the thoracic spine, especially T₆ and T₄. Pressure paraspinous to these levels elicit buttocks' pain.

raising may be painful and some sensory or motor deficit can be elicited.

Entrapment of the lateral femoral cutaneous nerve, termed meralgia paresthetica, causes a burning type of pain in the anterolateral portion of the thigh (Fig. 181). Usually the area is very clearly delineated and ultimately the pain may be accompanied by numbness of the area. The lateral femoral cutaneous nerve (L_2 to L_3) lies within the pelvis and emerges superficially in the region of the anterosuperior iliac spine to lie beneath the deep lateral fascia of the thigh. Entrapment usually is at the lateral end of the inguinal ligament below the anterosuperior spine. Often the history of trauma cannot be elicited.

Treatment is expectant. Posture exercises to decrease lordosis are considered to be valuable. Often applying a heel lift on the opposite shoe relieves tension upon the fascia lata and affords relief. Oral anti-inflammatory medicine should be taken. Surgical decompression of the nerve is rarely indicated.

214

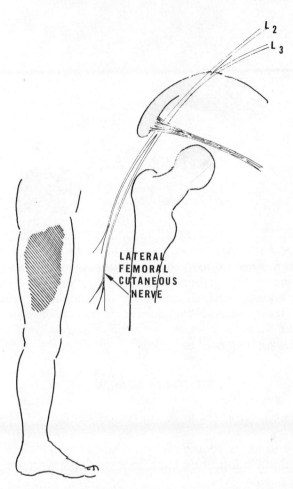

FIGURE 181. Meralgia parenthetica. Irritation of the lateral femoral cutaneous nerve anywhere along its course from L₂ to L₃ distally can cause a burning pain in the cutaneous zone depicted.

BIBLIOGRAPHY

Blount, W. P.: Don't throw away the cane. J. Bone Joint Surg. 38[Am]:695, 1956.

Cailliet, R.: Foot and Ankle Pain. F. A. Davis Co., Philadelphia, 1968.

Denham, R. A.: Hip Mechanics. J. Bone Joint Surg. 41[Br]:550–557, 1959.

Gardner, E.: The innervation of the hip joint. Anat. Rec. 101:353–371, 1948.

Inman, V. T.: Functional aspects of the abductor muscles of the hip. J. Bone Joint Surg. 29:607, 1947.

Lloyd-Roberts, G. C.: The role of capsular changes in osteoarthritis of the hip. J. Bone Joint Surg. 37[Br]: 8–47, 1955.

Saunders, J. B., Inman, V. T. and Eberhart, H. D.: The major determinants of normal and pathological gait. J. Bone Joint Surg. 35[Am]:543–558, 1953.

Trueta, J., and Harrison, M. H. M.: Normal vascular anatomy of femoral head in adult man. J. Bone Joint Surg. 35[Br]:442–461, 1953.

Wadsworth, J. B., Smidt, G. L., and Johnston, R. C.: Gait characteristics of subjects with hip disease. Phys. Ther. 52:829–837, 1972.

CHAPTER 10

Knee Pain

The knee probably is the most vulnerable joint in the body to become a source of pain. This is due to the joint's importance in gait and stance and, more important, in bending, stooping, and squatting. By its structure the knee is unstable. It is totally dependent upon ligamentous support and strong muscular function. It has an extensive synovial membrane. All these factors have been fully discussed in *Knee Pain and Disability*,[1] but are beneficial to review.

STRUCTURAL ANATOMY

There are two joints in the knee: the femorotibial and the femoropatellar (Fig. 182).

The distal end of the femur has two convex condyles separated by a deep V-shaped notch inferiorly and a concave depression anteriorly into which fits the patella.

The femoral condyles articulate with the articular concave surface of the tibial plateau. This forms the femorotibial joint. The articular surfaces are not symmetric so they do not form a stable congruent joint. Symmetry is created by the interposition of fibrocartilaginous menisci that assist in distributing pressure between the femur and the tibia, increase the elasticity of the joint, and assist in its lubrication.

Ligaments

There are strong ligaments on the medial and lateral aspect of the joint extending from the femoral condyles to the tibia and fibula (Fig. 183); these give the major support to the joint. In the centrum of the tibiofemoral joint are the cruciate ligaments that add to the stability and assist in the normal mechanical function of the knee.

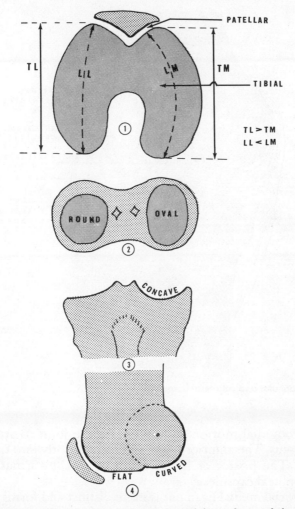

FIGURE 182. Knee joint surfaces. *1*, Femoral condyle surfaces of the right knee. *TL*, Anteroposterior length of the lateral condyle; *TM*, Length of the medial condyle. The length of the medial condyle, *LM*, is greater than the length of the lateral condyle, *LL*, because of its curved surface. *2*, Superior surface of the right tibia. The lateral articular surface is rounded and the medial articular surface is oval. *3*, The medial tibial articular surface is deeper and more concave than is the lateral. *4*, Side view of the femur showing the flat anterior surface and the curved posterior surface. The two articulations are illustrated in *1*: the patellar in which the patella articulates with the anterior femur and the tibial then glides upon the tibia.

The fibrous capsule of the joint has selective thickenings which form ligaments.

MEDIAL CAPSULAR LIGAMENTS. The medial capsular ligaments divide into deep and superficial portions (Fig. 184). The deep portion is also divided into three sections. The middle fibers stabilize the joint

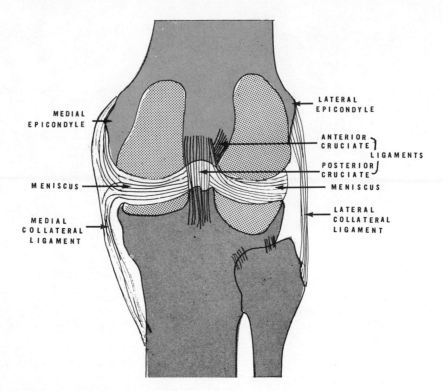

MEDIAL
EPICONDYLE

LATERAL
EPICONDYLE

ANTERIOR
CRUCIATE ⎤
⎥ LIGAMENTS
POSTERIOR ⎦
CRUCIATE

MENISCUS

MENISCUS

MEDIAL
COLLATERAL
LIGAMENT

LATERAL
COLLATERAL
LIGAMENT

FIGURE 183.   Capsular and collateral ligaments.

against lateromedial motion and penetrate the joint to attach to the medial meniscus. The anterior fibers extend anteriorly into the extensor mechanism. The posterior fibers extend into the formation of the posterior popliteal capsule.

The superficial medial ligament is more distinct and forms the medial collateral ligament. It is attached superiorly to the medial femoral epicondyle and inferiorly upon the tibia just below the level of the articular cartilage. There are numerous bursae between the deep and the superficial capsular ligaments.

LATERAL CAPSULAR LIGAMENTS. The fibular collateral ligament, a distinct thickening of the capsule, passes from the lateral epicondyle of the femur to the head of the fibula. In its course it is surrounded by the divided tendons of the biceps and is penetrated by the popliteus tendon as the tendon passes to attach to the lateral epicondyle of the femur. The peroneal nerve passes the neck of the fibula behind the biceps tendon in the region.

The knee has its maximum stability at full extension when the collateral ligaments are taut. Immediately upon flexion the collateral

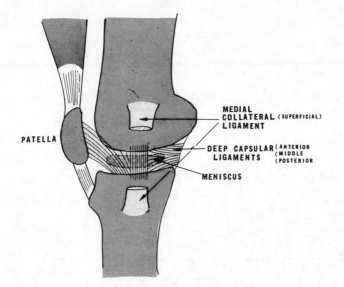

FIGURE 184. Medial capsular ligaments. The superficial collateral ligament attaches superior to the medial femoral condyle and attaches onto the tibia below the articular cartilage. The deep capsular ligament divides into three portions: anterior, middle, and posterior.

ligaments relax and permit lateromedial motion and rotation of the tibial upon the femoral condyles.

CRUCIATE LIGAMENTS. The paired cruciate ligaments are named according to their tibial attachment (Fig. 185). The anterior ligament originates from the anterior tibial plateau and proceeds superiorly and posteriorly to attach to the medial aspect of the lateral femoral condyle. The posterior ligament arises from the posterior aspect of the tibia and extends forward, upward, and inward to attach to the medial femoral condyle.

By their attachments and direction they restrict shear motion of the joint and thus act in flexion-extension of the knee. The anterior cruciate prevents knee hyperextension and the posterior cruciate mechanically assists the knee in flexion (Fig. 186).

Menisci

The menisci are curved, wedged, fibrocartilaginous structures that lie between the femoral condyles and the tibial plateau. The medial meniscus is approximately 10 mm. wide with its posterior horn wider than the anterior and middle portion. The medial meniscus forms a wider curve (C curve) than the lateral meniscus which is rounder and

219

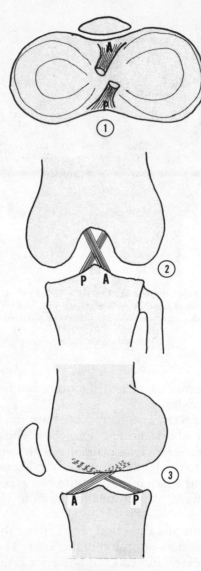

FIGURE 185. *1,* Superior view of the tibial plateau; *2,* Anterior view with knee extended; *3,* Lateral view. (*A,* Anterior ligament; *L,* Lateral ligament).

shaped like an O. Their outer margins are thicker and taper toward the centrum.

The medial meniscus is attached around its entire periphery to the joint capsule and the medial collateral ligament (Fig. 187). Its anterior horn connects to the anterior intercondylar eminence, to the anterior cruciate ligament, and via the ligamentous transverses to the lateral

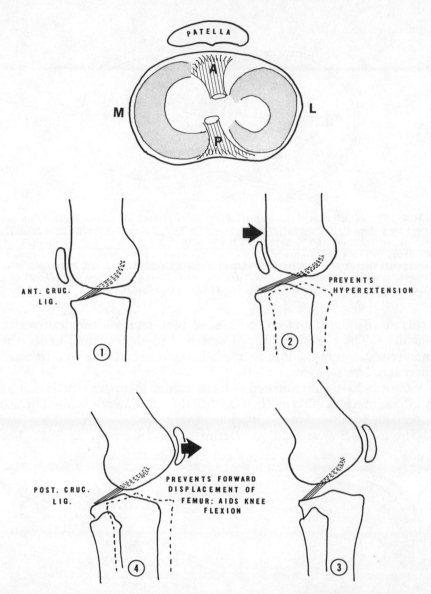

FIGURE 186. Cruciate ligament function. *Above*, Superior view showing the cruciate origination and direction. *1* and *2*, Lateral view showing anterior cruciate as it prevents hyperextension. *3* and *4*, Posterior cruciate ligament function, which prevents lateral displacement of the tibia upon the femur and aids in normal knee flexion.

meniscus. By virtue of these connections the medial meniscus moves with the tibia and femur and is exposed to mechanical trauma.

The lateral meniscus, 12 to 13 mm. wide, has the anterior and posterior horns attached directly to the intercondylar eminences, to the posterior

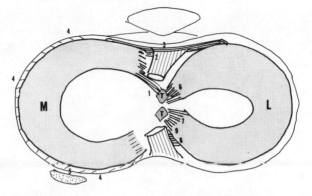

FIGURE 187. Attachment of the menisci. Right tibial plateau view from above. *1*, Fibrous attachment of medial meniscus *(M)* to tibial tubercle *(T)*. *2*, Connection to anterior cruciate; *3*, Transverse ligament which connects to the anterior horn of the lateral meniscus *(L)*. *4*, Meniscus *(M)* is attached around the entire periphery to the capsule. *5*, Attached to semimembranosus muscle tendon. *6*, Lateral meniscus anterior horn and *7*, posterior horn attached to eminentia intercondylaris *(T)* and attached to posterior cruciate ligament *(8)*. *9*, A fibrous band attaches superiorly into the fossa intercondylaris at the femur.

cruciate ligament, and to the medial meniscus via the transversus ligament. The periphery is not attached to the capsule, hence the meniscus can rotate about its medial attachment and is free to avoid mechanical entrapment.

Statistics have substantiated that the medial meniscus sustains injury 3:1 (to as much as 20:1) more often than does the lateral meniscus which is partially attributed to the attachment variation.

The menisci have a unique intrinsic blood supply (Fig. 188). The

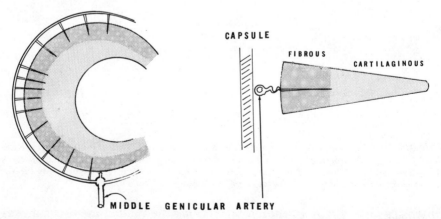

FIGURE 188. Intrinsic circulation of the menisci. The middle genicular branch of the popliteal artery sends branches around the periphery of the meniscus under the capsule. Small tortuous nonanastomatic vessels enter the outer fibrous zone of the meniscus. Their tortuousity permits movement of the meniscus. The inner third of the meniscus is cartilaginous and avascular.

222

middle genicular artery, a branch of the popliteal artery, branches circuitously around the menisci and sends small tortuous vessels into the outer third of the menisci. The middle and inner third of the menisci are avascular. This vascular factor accounts for repair of meniscus injuries to the outer third (fibrous portion) and the failure to heal in the inner two thirds (avascular cartilaginous portion).

Nerves

The knee joint has a rich sensory innervation that can transmit pain. All these tissues are innervated by the same nerves: the skin, the synovial membrane, the capsule, the ligaments, the muscles, and the bursae.

The skin is supplied primarily by the femoral and the obturator nerves (Fig. 189). There is a minor supply by the sciatic nerve. The synovial capsule is a relatively insensitive tissue and the articular cartilage carries no sensory fibers.

The fibrous capsule and the ligaments are richly supplied by medulated

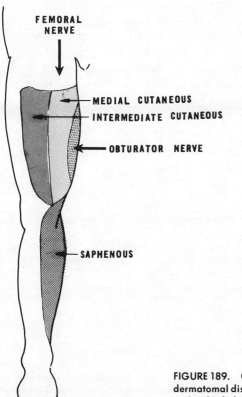

FEMORAL NERVE

MEDIAL CUTANEOUS

INTERMEDIATE CUTANEOUS

OBTURATOR NERVE

SAPHENOUS

FIGURE 189. Cutaneous regions of femoral nerve. The dermatomal distribution of the femoral nerve is depicted in the shaded areas.

223

and nonmedulated afferent somatic nerves capable of carrying pain sensation. Some of these articular nerve fibers penetrate the synovial membrane and can elicit pain from these regions. The capsular and ligamentous structures are innervated by the sciatic nerve (articular branch to the lateroposterior area). The tibial articular branch supplies the posterior aspect of the joint and the external popliteal nerve supplies the lateral articular area. The obturator nerve also sends a small branch to the posterior capsule. The anteromedial aspect of the capsule is supplied by the femoral nerve (Fig. 190).

The arterioles of the synovium are supplied by autonomic fibers and have sensory somatic fibers. Thus vascular changes initiate somatic changes; this can explain marked painful reaction claimed by arthritic patients to heat, cold, or barometric pressure changes.

Muscles

The knee joint is stabilized and powerfully motored by muscles that cross the joint from origin above the hip joint and from the shaft of the femur to insert upon bony structures below the knee joint. These muscle groups are commonly classified as extensors (anterior), flexors (posterior), adductors (medial), and abductors (lateral). The abductors and adductors only act upon the knee as rotators when the joint is flexed.

The extensors that are of greatest importance in the stability and function of the knee are the quadriceps femoris muscles. They are composed of four heads: the rectus femoris and three vasti, which are termed medialis, lateralis, and intermedius. The rectus originates from the anterior iliac spine and all the vasti from the shaft of the femur. All

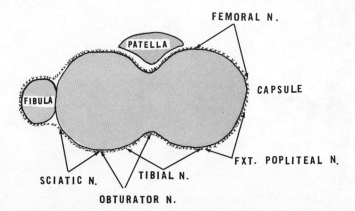

FIGURE 190. Cutaneous nerve distribution of knee area. Viewed from above, the knee capsule receives skin innervation from the sciatic, obturator, tibial, femoral, and external popliteal nerves. Pain can be elicited from capsular irritation at any of these areas.

four heads converge into a common tendon that crosses the knee joint to attach to the tibial tubercle.

Within the patellar tendon lies the patella which provides mechanical leverage to the extensor mechanism and provides a gliding surface against the femur to minimize friction. The quadriceps is innervated by the femoral nerve formed by the anterior division of L2, L3, and L4.

Besides extending the tibia upon the femur the extensor mechanism also has ligamentous attachments to the menisci (Fig. 191) that permit the menisci to move during knee motion to prevent entrapment.

The flexors are in the posterior aspect of the femur (Fig. 192). They cross the knee and flex the leg upon the thigh and rotate it. The flexors are best divided into medial and lateral groups with the medial containing the semimembranous, semitendinous, and the lateral biceps muscle of the thigh. With the knee flexed, the medial group rotates the leg internally and the biceps rotates the leg externally.

All the flexors originate from a common site on the ischial tuberosity. The semitendinous muscle descends the medial aspect of the thigh and, as it crosses the knee, it joins the sartorius and gracilis muscles to form a common tendon, the pes anserinus (Fig. 193). This tendon flexes the knee and internally rotates the flexed leg. A bursa separates this tendon

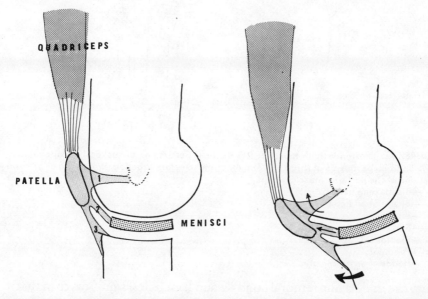

FIGURE 191. Quadriceps mechanism. The quadriceps extends over the anterior knee joint with three ligamentous extensions: 1, the epicondylopatellar portion attaches to the epicondyle eminence of the femur and guides rotation of the patella; 2, the meniscopatellar attaches to and pulls the meniscus forward during knee extension; and 3, the infrapatellar tendon, which attaches to the tibial tubercle and extends the tibia upon the femur.

225

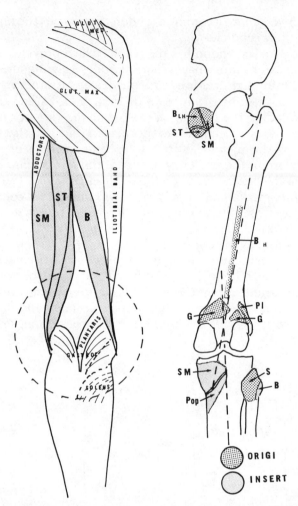

FIGURE 192. Posterior thigh muscles: flexors. *Left,* Semimembranous *(SM);* semitendinous, *(ST);* and biceps muscle of thigh *(B).* The other muscles are labelled. *Right,* The origin and insertion of the posterior muscle groups.

B_{LH} —Biceps long head
B_{SH} —Biceps short head
B —Biceps
S —Sartorius
PI —Plantar
Pop—Popliteal
G —Heads of gastrocnemius

from the underlying femoral condyle and has clinical significance in that it can cause pain.

The semimembranous insertion divides into four tendons that blend into the capsule, but it has a deep fibrous extensor that attaches to the medial meniscus to pull it posteriorly as the knee flexes (Fig. 194).

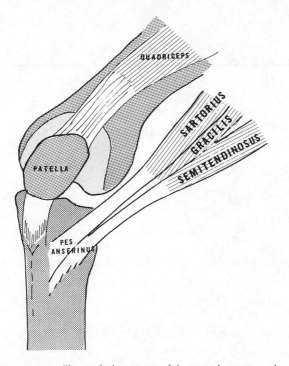

FIGURE 193.   Pes anserinus. The medial insertion of the outer hamstring, the semitendinous, forms a conjoined tendon with the sartorius and gracilis muscles to form the pes anserinus. A bursa is interposed between the tendon and the femoral condyle.

The biceps femoris tendon attaches to the head of the fibula by three fibrous insertions. One insertion flexes the knee and externally rotates the leg. The middle layer of the biceps tendon pulls upon the collateral ligament and causes it to bow and thus slacken. This deep portion also attaches to the capsule and during knee flexion prevents its being impinged between the tibia and the femur.

The flexor groups receive their innervation from the sciatic nerve. As the sciatic nerve divides into the tibial and the common peroneal nerve the former innervates the semimembranous, semitendinous, and long head of the biceps and the common peroneal nerve, the short head of the biceps.

The synovial capsule of the knee joint is large (Fig. 195), holding up to 40 cc. of air before it is distended. Anteriorly it ascends to two finger breadths above the patella; posteriorly it ascends to the origin of the gastrocnemius muscle; laterally it extends superiorly to the margin of the epicondyles and inferiorly to 1 cm. below the articular margin where the collateral ligaments attach.

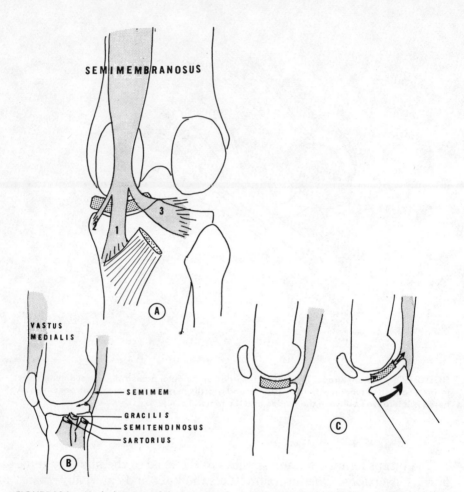

FIGURE 194. Medial aspect of the posterior knee structure. A, The semimembranous muscle has four tendinous inserts. The major insert, 1, extends to attach on the posterior aspect of the tibia and sends fibers into the popliteus. In its path there is an exterior branch that attaches to the posterior aspect of the medial meniscus, 2 and 3. These tendons complete the posterior popliteal fossa and tense the capsule. B, The medial aspect of the knee with the insertion sites of the medial flexors. C, The semimembranous flexes the knee and simultaneously pulls the meniscus backwards and rotates it with the tibia.

FUNCTIONAL ANATOMY OF THE KNEE

The knee flexes and extends in an intricate manner which predisposes the knee to damaging stresses when violated. Knee flexion and extension is a gliding movement of the tibia upon the femur with simultaneous rotation. As the knee flexes the lower leg internally rotates upon the femur and during extension externally rotates.

228

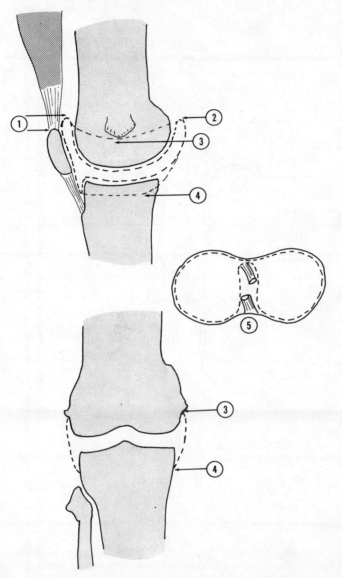

FIGURE 195. Synovial capsule. The capsule anteriorly ascends to two finger breadths above the patella *(1)*; posteriorly it ascends to the origin of the gastrocnemius muscle *(2)*; laterally *(3)* it attaches to the femur at the epidondylar level *(3)*; inferiorly it attaches upon the tibia ¼ inch below the articular margin just below attachment of the collateral ligament *(4)*. The cruciate ligaments *(5)* invaginate the capsule but are extracapsular.

Because of the contour of the femoral head with its anterior surface flat and the posterior portion curved, the first 20 degrees of flexion is that of a rocking motion followed by gliding until the tibial surface rotates about the posterior curved femoral condyles (Fig. 196). Further flexion is

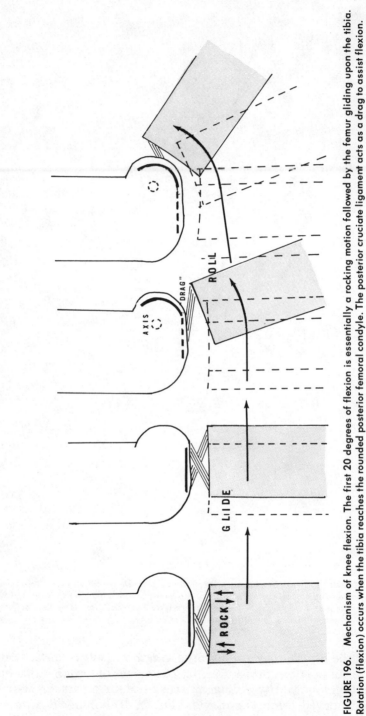

FIGURE 196.   Mechanism of knee flexion. The first 20 degrees of flexion is essentially a rocking motion followed by the femur gliding upon the tibia. Rotation (flexion) occurs when the tibia reaches the rounded posterior femoral condyle. The posterior cruciate ligament acts as a drag to assist flexion.

assisted by the posterior cruciate ligaments which, once fully elongated, create a fulcrum about which the tibia moves.

Once flexion begins, the capsular ligaments relax to permit rotation. Most rotation occurs during the final phases of full flexion and full extension, but some degree of rotation occurs throughout flexion and extension. With the knee flexed to 90 degrees the lower leg can rotate 40 degrees. With the knee fully extended the joint surfaces are in direct opposition and the collateral ligaments taut so that no rotation or no lateral deviation of adduction or abduction is possible. The fully extended knee is stable and resistant to stress.

The direction and extent of rotation is dependent upon the anatomic configuration of the articular surfaces. The medial femoral condyles is longer in its curved axis than is the lateral condyle so as extension or flexion upon the lateral articular surface is reached there is some remaining gliding surface upon the medial condyle, hence external rotation into further extension and internal rotation into further flexion (Fig. 197).

Muscular action also assists rotation at the knee. The quadriceps group oblique medially across the knee joint and rotates the tibia externally during extension. The popliteal muscle begins internal rotation of the tibial during initiation of flexion.

The menisci are fixed to the tibia and femur and thus move during knee flexion and extension. In flexion-extension the menisci move with the tibia whereas in rotation the menisci move with the femur upon the tibia. The menisci are not subject to impingement if the flexion-extension motion pattern is not violated.

The cruciate ligaments, which assist in flexion (the posterior cruciate), also crisscross and thus limit rotation of the tibia upon the femur. The posterior cruciate ligament prevents excessive internal rotation and the anterior cruciate ligament prevents excessive external rotation. It is evident that abnormal shearing forces or rotational forces upon the knee can cause disruption of the cruciate ligament as well as damage the menisci once the capsular ligaments are overstretched or torn.

PATELLOFEMORAL ARTICULATION

The part of the femoral surface that articulates with the patella is saddle shaped and asymmetric. The lateral face is larger and more convex than the medial surface. The patella has a smooth posterior articulation surface covered with cartilage presenting two facets separated by a vertical ridge. This ridge fits into a corresponding groove between the two condyles of the femur (Fig. 198).

The stability of the patella in remaining within its articulation is dependent upon the depth of the groove, the proper contour of the

231

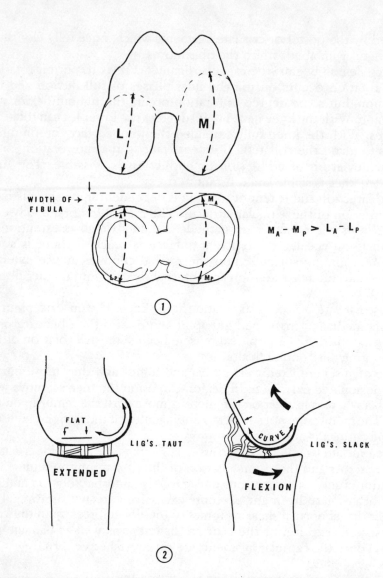

FIGURE 197. Passive mechanism of knee flexion-extension. *1*, The articular surface of the medial condyle is longer than that of the lateral (M_A to M_P greater than L_A to L_P); thus the tibia travels further upon the medial condyle during extension and rotates externally. *2*, Because of the flat anterior femoral surface the collateral ligaments are taut and the knee is stable. They relax upon flexion.

patella, and adequate muscular mechanism. The low medial attachment of the vastus medialis muscle to the patella aids in preventing lateral subluxation.

The patella moves along the femoral condyle during flexion and extension and both aids and hinders mechanical efficiency. In the last 30

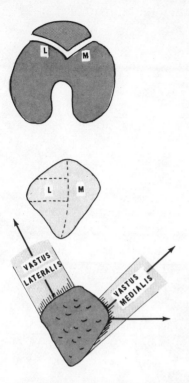

FIGURE 198. Patella. *Above,* Patellofemoral articulation. The lateral condyles are broader and more concave than the medial. *Center, Facet* planes of the patella on the articular surface: three planes on the lateral half and one plane on the medial. *Below,* Muscle pull upon the patella with the vastus lateralis muscle pulling cephalad. The vastus medialis muscle, by attaching lower and more laterally, exerts medial pull, thus centralizing and stabilizing the patella.

to 40 degrees of extension the patella adds to the efficacy of the extensor mechanism. In the fully flexed knee the presence of the patella hinders extension.

The alignment of the extensor mechanism and the length of the infrapatellar ligament have recently been considered of importance in the production of patellar chondromalacia. An increase in the Q angle of more than 20 degrees or an increased length of the patella have been considered conducive to chondromalacia (Fig. 199).

PAIN SYNDROMES

Complaints of pain in the knee must be clarified by clinical manifestations. Whereas the predominant low back and cervical pain diagnoses are made by history, knee pain requires a careful examination. Pain causes the patient to seek help and localizes the site, but a careful examination is mandatory.

Steindler[2] attempted to clarify pain localized anteriorly, anteromedially, and anterolaterally in the knee into specific structures involved (Fig. 200). He specified the site of trigger points, points of maximum tenderness, as being valuable in differentiation of the specific structure involved.

233

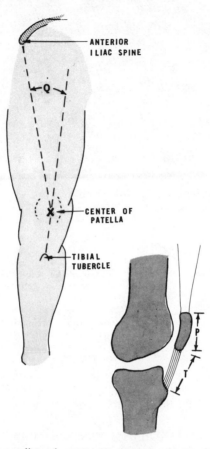

FIGURE 199. Q angle—patellotendon ratio. The Q angle is drawn from a line originating at the anterior iliac spine to the center of the patella (essentially the direction of pull of the quadriceps) and a line from the center of the tibial tubercle to the center of the patella. Normal is considered to be 20 degrees or less. Roentgenographic measurement of the length of the patella *(P)* compared to length of infrapatellar ligament *(T)* determines the height the patella rides. T should equal P. If T is longer it indicates a high-riding patella.

Fat Pad: Medial or Lateral

There are fat pads in the knee joint that are intracapsular, albeit extrasynovial. They alter their shape with movement of the joint. Fat pads are seen especially in weight-bearing lower extremities of mammals. The fat pads of the knee are closely packed fat cells contained in a considerable amount of elastic tissue. It thus makes a firm pad that is deformable when subjected to pressure and regains its shape upon release.

Fat pads of joints are liberally supplied with pain receptor-type nerve

234

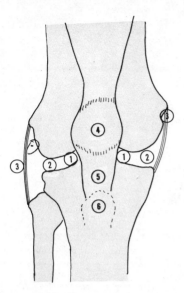

FIGURE 200. Tender sites of anterior aspect of the knee. *1*, The site of painful fat pads; *2*, Meniscus sites of tenderness; *3*, Collateral ligament pain (medial and lateral); *4*, Patellar pain and tenderness (see text for explanation); *5*, Infrapatellar bursal pain; *6*, Tibial tubercle (Osgood-Schlatter's disease).

endings. Their function is considered to be to fill dead spaces and assist in lubrication.

If the fat pad is abnormally large, if the quadriceps mechanism is relaxed, or if the joint is used in a faulty manner the pad can be impinged with internal hemorrhage. Pain results and frequently the patient claims that the knee "gives out." Examination may reveal a hypertrophied pad. Tenderness is felt just medial (lateral) to the patellar tendon. (See Fig. 200–1.) Forced passive extension of the knee increases the pain. The knee may lock, but this is unusual. Bilateral fat pads, in the same knee, occur more often when there is concomitant degenerative arthritis.

Injection of the pad may be diagnostic and momentarily therapeutic, but usually surgical removal is indicated.

Meniscus Injury

As previously stated, the medial meniscus sustains injury more often than the lateral (3:1, 8:1, to 20:1) depending on the authority.

The mechanisms causing meniscus injury are numerous but are predominantly compression and/or traction. The usual injury mechanism is rotatory stress on the weight-bearing leg. The stress is imposed by violation of internal rotation during flexion or external rotation during extension. These motion patterns are physiologic.

During knee flexion and extension the menisci move anteriorly and posteriorly respectively. With maximum flexion the posterior portion of the meniscus is compressed between the posterior aspect of the tibial and femoral condyles. Rotation of the tibia upon the femur in this flexed

235

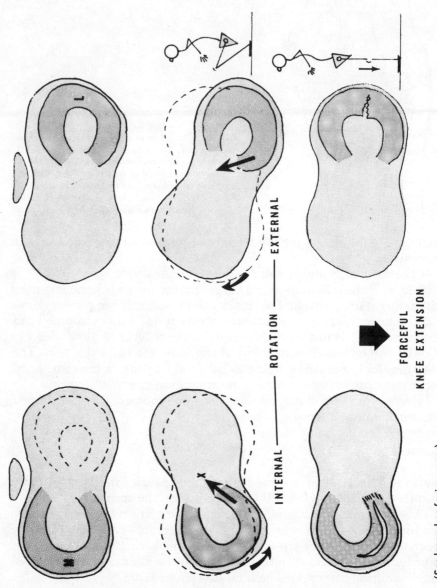

ROTATION — EXTERNAL

ROTATION — INTERNAL

FORCEFUL KNEE EXTENSION

FIGURE 201. (See legend on facing page.)

position displaces the posterior horn of the meniscus towards the center of the joint. Now forceful extension of the knee either tears the posterior attachment or causes a longitudinal tear in the meniscus (Fig. 201). The direction of rotation decides which meniscus is entrapped and what type of tear occurs (Fig. 202).

Tearing of a meniscus with the knee fully extended is rare unless the stress is violent and disrupts the collateral or cruciate ligaments with or without condylar fracture. Meniscus injury requires flexion and extension of the knee combined with inappropriate rotation when the lower leg is fixed to the ground in a weight-bearing position.

Another mechanism considered to cause injury to menisci is that of a forced valgus of the knee during flexion and external rotation that excessively, albeit momentarily, opens the joint space and, consequently, entraps the meniscus.

A complete longitudinal tear at the initial injury is considered rare. Complete tearing is considered to occur from repeated injuries.

The medial meniscus sustains its initial tear most commonly in the posterior pole and a longitudinal tear occurs usually in the posterior third of the meniscus. If the tear extends anteriorly past the collateral ligament it bunches up between the condyles and can cause joint locking. Extensive tear may result in the fragment protruding into the center of the joint with *no* resultant locking.

The resultant effusion of a meniscus tear is from injury to the synovium. A large effusion can cause pain and also limited motion.

Other causative factors are implicated. They include constitutional inadequacy, ligamentous laxity, faulty work habits, excessive knee varus or valgus, and violent sports.

DIAGNOSIS OF MEDIAL MENISCUS INJURY. Diagnosis of medial meniscus injury is made by careful evaluation of the history which describes the mechanism. The history must also include previous similar episodes.

Usually pain is severe and sudden and causes immediate cessation of activities. Locking may occur but may not be immediate; it may occur hours later and may be transient. Usually effusion is present following the injury. The lateral meniscus, being more loosely attached to the capsule, causes less effusion than does the medial meniscus.

When the anterior peripheral meniscus attachment is involved, the tender spot (trigger area) is between the site of the fat pads and the

---

FIGURE 201. Mechanism of meniscus tear. *Center,* With the knee flexed the posterior horns (X) move toward the center of the joint. Dependent on the direction of rotation either the medial or lateral meniscus so moves. *Below,* As the person extends the knee the meniscus can sustain a tear.

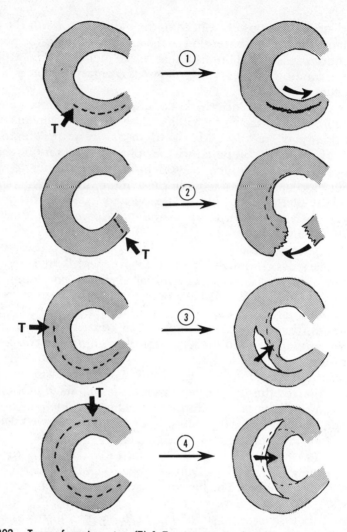

FIGURE 202. Types of meniscus tear (T). 1, Tear in posterior third of meniscus. Because of the elasticity of the cartilage the meniscus springs back into its normal position. 2, Posterior tear with avulsion causes the anterior portion of the meniscus to bunch. 3, Longitudinal partial tear may cause bunching of inner third of meniscus with resultant locking. 4, Complete longitudinal bucket handle tear. The central portion of the meniscus moves into the center of the joint away from the condyles and thus the joint does not lock.

collateral ligaments. (See Fig. 200–2.) The tender area in injury to the posterior attachment lies behind the collateral ligament and may be concealed behind the hamstring muscles. It is difficult to elicit.

When the meniscus injury is midpoint the tenderness is at the site of the collateral ligament and makes it difficult to differentiate from ligamentous strain or tear. Other diagnostic signs are necessary for differentiation.

Frequently pain is referred to the contralateral side from the injury. A lateral meniscal tear may refer pain to the medial side. The converse is rare. Tenderness on the medial side usually implicates the medial meniscus.

LATERAL MENISCUS TEARS. Lateral meniscus tears also occur, although there are fewer than medial meniscus tears. Lateral meniscus tears can refer pain to the medial side of the knee.

Lateral tears usually occur in the posterior portion; hence the tender (trigger) area is posterolateral behind the fibular collateral ligament and is made worse by *flexion* of the knee. Bucket handle variety is rare. Effusion is usually less and may be absent. Because of its unique attachments the lateral muscle meniscus has the propensity to reduce itself automatically.

Meniscal Cysts. Cysts can develop in degenerated menisci and usually occur in the lateral rather than the medial menisci. These may be painful and are difficult to differentiate clinically from tears. The cysts protrude when the knee is extended or fully flexed but recede between these two excessive positions. The knee is most comfortable when it is flexed at 130 to 120 degrees. The triad of meniscal cyst is pain, interference with joint motion (full extension and full flexion), and a palpable tender tumor that recedes with slight degree of flexion.

DIAGNOSTIC MENISCUS SIGNS. Meniscus signs (Fig. 203), also termed as signs of internal derangement of the knee, are numerous and unfortunately are confusing insofar as they are termed by the proper name of the physician who originally described that specific maneuver—McMurray's sign, Apley test, and hyperflexion test (described in *Knee Pain and Disability*).[3] All attempt to reproduce the symptoms or confirm the mechanical impairment by flexing or extending the knee with various degrees of rotation.

Collateral Ligament Pain Syndromes

A similar force or stress that causes meniscal tears can cause ligamentous tears. The latter usually are avulsions from the bony attachment with the tender (trigger) areas usually at their superior attachment, that is, above the joint line. (See Fig. 200–3.) The injury may be a lateral force with no rotational component (extreme valgus or varus).

In a strain without significant avulsion of the ligament the knee does not become unstable. When tear occurs, excessive motion of abduction or adduction (*with the knee fully extended*) of the lower leg upon the femur can be elicited. Swelling over the tear site may be noted. Stability of the opposite knee must be compared to ascertain the degree of mobility.

Ligamentous injuries may be accompanied by meniscus injuries and

239

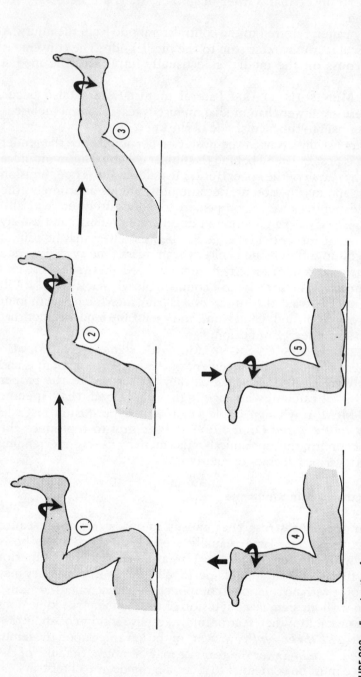

FIGURE 203. Examination for meniscus signs. 1, 2 and 3, Stages of McMurray test. With the patient supine the knee is fully flexed. The leg is internally rotated (lateral meniscus test) or externally rotated (medial meniscus test) and then the knee is fully extended. In a meniscus lesion a painful click is elicited which is most significant in the first phase of extension. 4 and 5, Apley test. With patient prone and leg rotated upward, pain upon traction (4) implies a capsular ligamentous lesion, while pain upon downward pressure (5) indicates a meniscus lesion.

this fact must always be kept in mind. The lateral collateral ligament is intimately related to the iliotibial band which can also be torn in a severe ligamentous avulsion.

Injury to the cruciate ligaments causes anteroposterior instability, but does not cause lateral instability per se. Suspicion of cruciate ligamentous tears can be verified by eliciting a positive drawer sign (Fig. 204).

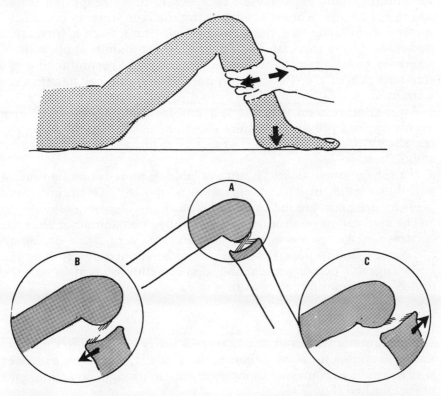

FIGURE 204. Drawer sign for cruciate ligament tears. *Above,* Position of patient's leg and examiner's hand. The foot is fixed upon the table and the lower leg is moved horizontally, proximally, and distally (arrows). *Below, A,* Intact cruciate ligaments; *B,* Posterior tibial movement in tear of posterior cruciate ligaments; *C,* Excessive anterior tibial motion in tear of anterior cruciate. If both cruciate ligaments are torn there is excessive motion in both directions.

TREATMENT OF LIGAMENTOUS AND MENISCAL INJURIES

When an injury is observed or upon sustaining a knee injury, the prevention of swelling is indicated immediately. Since swelling occurs from either microscopic or macroscopic hemorrhage and edema, vascular injury is implied as well as ligamentous and capsular injury. Curtailing the blood supply to the injured joint would appear feasible.

241

A sphygmomanometer should be applied proximal to the knee and inflated above systolic pressure. This prevents blood flow to the knee, thus minimizing swelling and further hemorrhage. It also permits a more leisurely and accurate examination and permits the application of ice and a pressure dressing before any swelling occurs.

Compression by a sphygmomanometer can be sustained safely for 15 minutes and then released completely to be reapplied for an additional period. Theoretically during the compression period the injured capillaries are permitted to clot and edema formation decreases. I have successfully performed this proximal application of arterial occlusion but in too few patients as yet to recommend it as a standard procedure. Further physiologic and clinical statistics are warranted.

After compression with the sphygmomanometer, ice and a firm evenly-applied pressure dressing should be applied to the knee, the leg should be elevated, and crutches made available to permit ambulation.

If a sphygmomanometer is not available or there is no one present with knowledge in its use, elevation of the leg, ice packing, and pressure dressings are indicated in its stead.

The application of a tourniquet or a sphygmomanometer after the occurrence of edema is not indicated and may actually be damaging. No constriction to the circulation should be applied unless there is a knowledgeable person present and someone who will remain with the injured person until the pressure is released.

Ligamentous Injuries

Ligamentous strains or minor tears actually heal well in time, and active exercises must be begun early. Initially, isometric exercises, which avoid joint motion but increase muscle strength and endurance, are prescribed (Fig. 205).

Isometric exercises are gradually followed by kinetic (isotonic) exercises graduating to progressive resistive exercises (Fig. 206).

If significant effusion occurs in the joint, aspiration should be performed early and repeated if effusion recurs (Fig. 207). Persistent effusion, especially if it is great and creates pressure and hemorrhage, causes mechanic and ischemic damage to the already-damaged, overstretched, and torn tissue.

Meniscal Injuries

Meniscal tears in the inner two thirds of the meniscus (the avascular zone) do not heal whereas those whose tears are in the outer vascular zone and are reduced can heal.

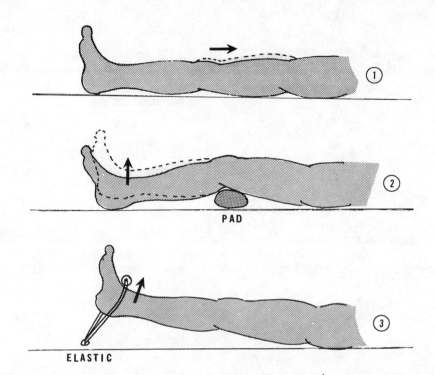

FIGURE 205. Types of quadriceps exercises. *1*, Quad setting. Knee is fully extended, muscle shortens with *no* joint motion, and patella ascends. *2*, with a 3 inch pad under the knee the quadriceps function is enhanced and joint motion is minimal. *3*, Straight leg raising against resistance (resisted isometric exercises) further strengthens quadriceps without joint motion.

If the displaced meniscus can be reduced and the leg immobilized for at least three weeks, recovery is possible. There are no specific indications for surgery, but the decision for surgery depends upon the surgeon's experience and expertise. As a rule, recurrent locking, sustained disability, or intractable pain are suggestive of arthrography and surgery.

Reduction must be initiated within 24 hours of locking. Reduction technique requires manual longitudinal traction with simultaneous rotation in both direction and lateral mobilization (into varus and valgus). The knee is placed in full flexion with knee forcefully internally rotated in a medial meniscus tear and externally rotated in a lateral meniscus tear. Then forceful kicking by the patient places the knee in full extension. If the attempt is unsuccessful, repeated manipulations or forceful attempts should not be considered.

After successful reduction and adequate immobilization, restoration exercises should be undertaken. The major exercise program should be to strengthen the quadriceps mechanism.

243

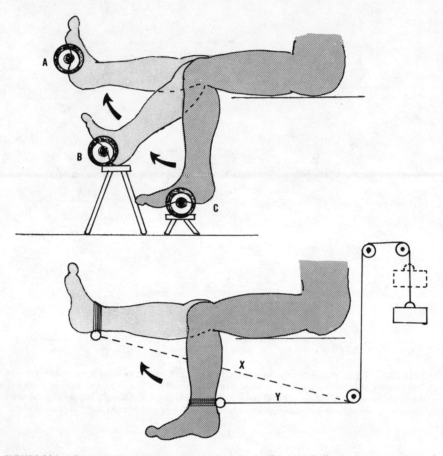

FIGURE 206.   Progressive resistive exercises (isotonic). *Above, A,* Full extension is attempted. *B,* with stool to support weighted foot, only the last degrees of extension are exercises. This is most desired and strengthens this vastus medialis. C, Resistance is minimal and there is ligamentous strain on the knee when the leg is fully dependent. *Below,* Pulley exercises. Maximum resistance occurs in first 45 degrees of extension and none at full extension.

EXTRA-ARTICULAR PAIN SYNDROMES

Housemaid's Knee

There is a bursa located beneath the skin and between the patella and the tibial tubercle that is subject to mechanical pressure during kneeling. Essentially there are two bursae in this region: one over the lower half of the patella and one over the superior half of the tibial tubercle (Fig. 208).

When the bursae are inflamed there is aching or pain on local

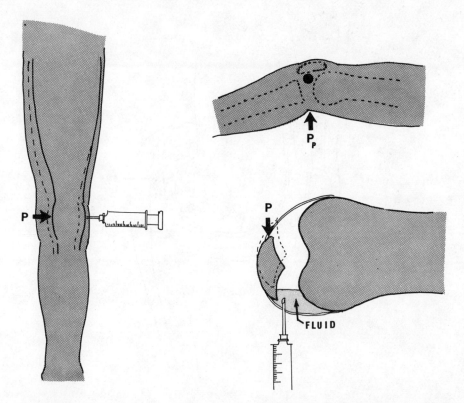

FIGURE 207. Knee aspiration technique. A, Patella (P) is moved laterally to increase injectable space between the patella and the femoral condyle. B, Lateral view of site of injection (black dot). Pressure upon popliteal space (P_P) brings fluid towards needle tip. C, In lateral position gravity localizes fluid to permit easier withdrawal.

pressure or upon kneeling. Swelling is usually noted and serous fluid can be aspirated.

Pes Anserinus Bursitis

The pes anserinus bursa is located superficially to the medial collateral ligament at the upper medial aspect of the tibia under the conjoined tendon of the sartorius, semimembranous, and gracilis muscles. Pain is increased by forceful extension of the knee or resisted contraction of the hamstrings in knee flexion.

Since there are numerous bursae about the knee joint, bursitis should be suspected in any local tenderness. Procaine injections combined with rest of the part are diagnostic and frequently therapeutic.

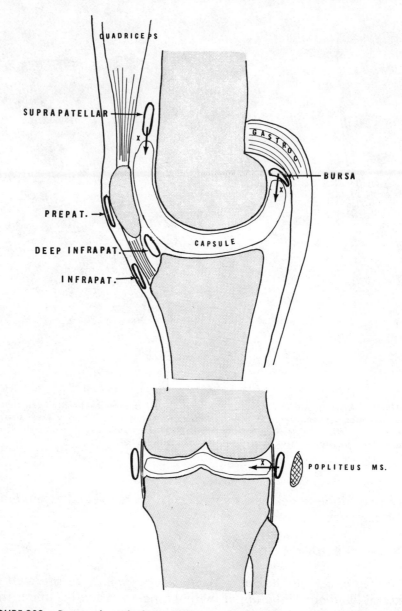

FIGURE 208. Bursae about the knee. The suprapatellar bursa may communicate with the knee capsule. X indicates other bursae that may communicate with joint space.

Baker's Cyst

First described by Baker in 1871, this cyst appears in the popliteal space. The cyst is lined by endothelium and communicates with the knee joint. Pain is more that of an ache or discomfort on walking and the cyst becomes visible and palpable. The protrusion may disappear on knee flexion and protrude on knee extension.

A Baker's cyst can be related to internal knee joint pathology which must be corrected before the cyst can be reduced. Excision of the cyst is no longer considered valid; synovectomy of the joint is more plausible. If the cyst causes inconvenience it can be excised.

DEGENERATIVE ARTHRITIS

The knee is a major site of degenerative arthritis. The initial symptoms are those of aching with some stiffness. Pain can be related to weight bearing but may also be characterized by stiffness from prolonged sitting. Crepitation may be noted on movement of the knee joint. Roentgenographic changes appear much later than do the clinical findings.

Treatment is both medical and physical. Salicylates in adequate prolonged doses remains the mainstay of medical treatment, but antiphlogistic drugs such as Indocin and Butazolidin have their advocates. Intra-articular injection of an anesthetic agent and steroids is valuable. Use of a cane is indicated on the side opposite the involved side. Crutches may be necessary when pain is severe. Body-weight reduction is imperative. The patient must remain active and ambulatory and must be instructed in exercises to maintain range of motion and quadriceps strength.

Daily activities must be evaluated. Low chairs must be avoided and prolonged sitting controlled. Deep knee bends should be avoided as should stair climbing and descending.

Patients with severe valgus and varus deformities are more prone to degenerative arthritis. This condition may require bracing or surgical intervention. Surgical procedures for relief of degenerative arthritis pain are beyond the scope of this book.

Rheumatoid arthritis is a systemic disease in which the knee is usually involved. The initial phase of joint disease is soft tissue involvement with effusion, heat, redness, pain, and limited motion. Diagnosis is made by history, general physical findings, laboratory test confirmation, synovial fluid examination, and ultimately roentgenographic findings.

Treatment has been well documented elsewhere[4] but it is well to emphasize that quadriceps exercises *must* be started early to avoid atrophy. Flexion contraction must be avoided, and oral medication and intra-articular injection of steroids after aspiration are valuable. (See Fig. 207.)

Treatment may ultimately involve surgical intervention.

Chondromalacia Patellae

Degenerative arthritis of the patellofemoral joint is as frequent and as painfully disabling as degenerative arthritis of the femorotibial joint.

There are numerous theories regarding the cause of this degenerative entity: abnormal ridges have been found on the femoral condyle, incongruity of the femoropatellar articulation, abnormal length of the infrapatella tendon, but most recently malalignment of the patella because of an increased Q angle (see Fig. 199) has been incriminated.

Symptoms are those of retropatellar aching, feeling of crepitation, a feeling of the knee "giving way" (but no locking), and pain in descending stairs or rising from a chair.

Examination reveals a normal knee but tenderness over the medial aspect of the patella. Pressure on the patella with simultaneous quadriceps contraction elicits crepitation and pain. Roentgenographic findings are negative until the condition is advanced. Chondromalacia patellae frequently is a complication of other knee lesions such as degenerative arthritis, cruciate or collateral ligament lesions, or even fat pad encroachment.

Nonsurgical treatment is often effective and consists of avoiding any irritating activity such as deep knee bends or squats, brief immobilization in a cylindrical cast, or retropatellar steroid injections.

Surgical intervention varies from arthrotomy and shaving the involved patella to total patellectomy. The current trend is combination of patellar realignment with excision of cartilage[5] that modifies the Q angle. In Europe the Maquet Principle is getting acceptance.

The Maquet Principle involves anterior displacement of the tibial attachment of the patellar tendon a distance of 2.5 cm. (Fig. 209). This is accomplished by performing an osteotomy and inserting a bone block. Changing the vector angle of the patellar tendon, the force of the patella against the femoral condyle is decreased as much as 50 percent. With this new position of the patella, the upper facet of the patella applies against the femur during knee flexion sooner than does the middle facet. Since this is a larger facet there is a relative decrease of surface weight-bearing area.

Lateral or medial patella deviation (Q angle) can be changed. This procedure is applicable for chondromalacia or recurring patellar dislocation.

OSTEOCHONDRITIS DISSECANS

Osteochondritis dissecans, which is a painful condition, is caused by a fragment of cartilage that becomes loose in the joint. It occurs in all ages

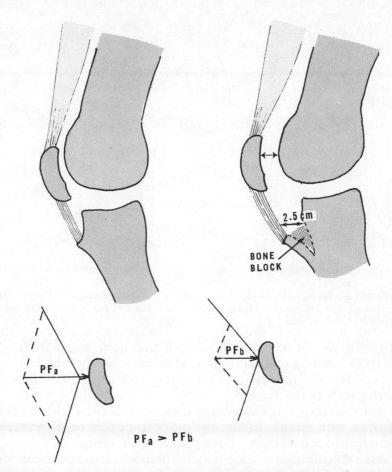

FIGURE 209. Anterior advancement of patellar tendon. *Left*, Normal patellofemoral relationship. *Right*, A 2.5 cm. advancement of the tibial tubercle attachment of the patellar tendon. The patella is further distant from the femoral condyle. The parallelograms reveal the patellar force *(PF)* to be less in the right figure.

but mostly in young males. The sites of the condition are indicated in Figure 210.

Onset usually is insidious with pain described as aching and with some stiffness. Depending upon the size and location of the fragment, locking can occur.

Diagnosis is confirmed by roentgenography. If time and periodic immobilization do not afford relief, surgical excision is indicated.

OSGOOD-SCHLATTER DISEASE

Osteochondritis of the tibial tubercle is called Osgood-Schlatter disease. It is observed mostly in adolescent boys. Complaint is of pain

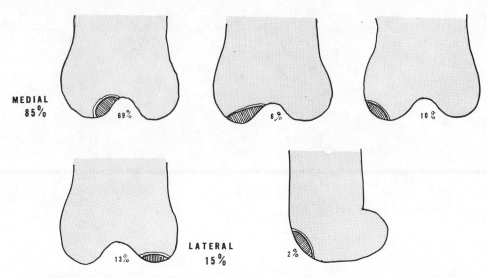

FIGURE 210. Sites of osteochondritis dissecans. Eighty-five percent occur on the medial femoral condyle with the classical site the lateral border of the medial condyle. Other sites and their percentages are listed.

and tenderness over the tibial tubercle and these are aggravated by kneeling. Swelling and tenderness is noted over the tibial tubercle. Etiologic factors are still questioned and roentgenography is usually negative early in the disease.

Treatment consists of avoiding excessive activities such as kneeling, squatting, or jumping. In more severe cases cylindrical casts can be used to prevent the youth from squatting or kneeling.

Larsen-Johannson's disease is a variant in which the osteochondritis affects the inferior pole of the patella. Symptoms and treatment are identical to Osgood-Schlatter disease and is identifiable by roentgenography.

REFERENCES

1. Cailliet, R.: Knee Pain and Disability. F. A. Davis Co., Philadelphia, 1973.
2. Steindler, A.: Lectures on the Interpretation of Pain in Orthopedic Practice. Charles C Thomas, Springfield, Il, 1959. Lecture XV, pp. 555–605.
3. Cailliet: Knee Pain and Disability, p. 47.
4. Cailliet: Knee Pain and Disability, pp. 92–96.
5. Insall, J., Falvo, K. A., and Wise, D. W.: Chondromalacia patellae. J. Bone Joint Surg. 58[Am]:1–8, 1976.

BIBLIOGRAPHY

Basmajian, J. V.: Grant's Method of Anatomy. Williams & Wilkins, Baltimore, 1971.
Basmajian, J. V., and Lovejoy, J. F.: Function of the popliteus muscle in man. J. Bone Joint Surg. 53[Am]:557, 1971.

Brantigan, O. C., and Voshell, A. F.: Tibial collateral ligament: Its function, its bursae, and its relation to the medial meniscus. J. Bone Joint Surg. 25:1, 1943.

DeLorme, T. L.: Restoration of muscle power by heavy resistance exercises. J. Bone Joint Surg. 27[Am]:645, 1945.

De Palma, A. F.: Diseases of the Knee: Management and Surgery. J. B. Lippincott Co., Philadelphia, 1954.

Helfet, A. J.: Mechanism of derangements of the medial semilunar cartilage and their management. J. Bone Joint Surg. 41[Br]:319, 1959.

Helfet, A. J.: Function of cruciate ligaments of knee-joint. Lancet 1:665, 1948.

Jack, E. A.: Posterior peripheral detachment of the lateral cartilage. J. Bone Joint Surg. 35[Br]:396, 1953.

Kuhns, J. G.: Changes in elastic adipose tissue. J. Bone Joint Surg. 31[Am]:541–547, 1949.

Last, R. J.: Some anatomical details of the knee joint. J. Bone Joint Surg. 30[Br]:683, 1948.

Lieb, F. J., and Perry, J.: Quadriceps function. An electromyographic study under isometric conditions. J. Bone Joint Surg. 53[Am]:749, 1971.

MacConall, M. A., Barnett, C. H., and Davies, D. V.: Synovial Joints: Their Structure and Mechanics. Charles C Thomas, Springfield Il, 1961.

Maquet, P.: Un traitment biomecanique de l'arthrose femero-patellaire. L'Avancement dei tendon ratalein. Rev. Rhumat. 30:779, 1963.

Marshal, J. L., Girgis, F. G., and Zelko, R. R.: The biceps femoris tendon and its functional significance. J. Bone Joint Surg. 54[Am]:1444, 1972.

McMurray, T. P.: The semilunar cartilage. Br. J. Surg. 29:407, 1942.

Murray, J. U. G.: The Maquet principle. Its application in severe chondromalacia patellar, patellofemoral and global knee osteoarthritis. Orthop. Review 8:29–36, 1976.

Ricklin, P., Ruttmann, A., and Del Buono, M. S.: Meniscus Lesions: Practical Problems of Clinical Diagnosis, Arthrography, and Therapy. Grune & Stratton, Inc., New York, 1971.

Slocum, D. B., and Larson, R. L.: Rotatory instability of the knee. J. Bone Joint Surg. 50[Am]:211, 1968.

Thorek, S. L.: Orthopaedics: Principles and Their Application, ed. 2. J. B. Lippincott Co., Philadelphia, 1967.

CHAPTER 11

Foot and Ankle Pain

The foot presents a unique segment of human anatomy in that all its parts are accessible to visual examination, direct palpation, and mechanical evaluation. The patient directs the attention of the examiner to the painful area or the disabling function and the examination should lead to a specific diagnosis. Complete knowledge of the functional anatomy is necessary in evaluation of foot disorders to an even greater degree than other portions of the musculoskeletal system. Meaningful treatment logically follows meaningful evaluation.

FUNCTIONAL ANATOMY

The foot is an intricate structure composed of 26 articulating bones constructed to bear full body weight and to transport the human body over varying types of terrain. The foot can be divided into three functional units: anterior, middle, and posterior (Fig. 211).

The posterior segment lies directly under the tibia and supports the superincumbent body. The talus in this segment is the mechanical keystone of the weight-bearing foot, articulating superiorly into the ankle mortice and inferiorly to the ground by contact with the calcaneus.

The talus has a body, a neck, and a head (Fig. 212). The superior and both sides of the body articulate with the tibia and fibula that unite to form the ankle mortice. The tibia contacts the entire superior aspect of the talus and is weight bearing. Its medial malleolus extends one third of the way down the medial aspect of the talus. The fibular malleolus covers the entire lateral aspect of the body of the talus. Within the mortice the talus functions as a hinge joint and permits dorsal and plantar motion of the foot at the ankle (Fig. 213).

Viewed from above the talus is wedge shaped with the anterior

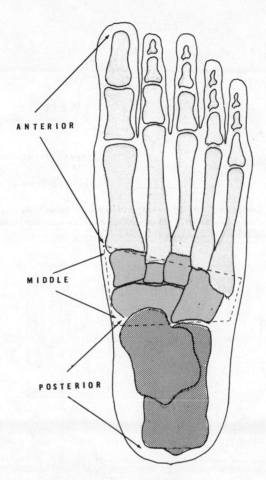

ANTERIOR

MIDDLE

POSTERIOR

FIGURE 211. The three functional segments of the foot.

portion being wider. As the ankle dorsiflexes, this wider anterior portion presents between the two malleoli of the mortice and wedges between them, separating the ankle mortice. No lateral or rotatory motion of the talus within the mortice is possible in this dorsiflexed position. As the ankle plantar flexes, the narrower posterior portion of the talus presents between the malleoli and allows mobility within the joint. This unstable position places added burden upon the ligaments which must bear the brunt of stabilizing the joint.

The talus moves in the mortice at an 18 degree angle since its axis of rotation is oblique to the coronal plane. The axis passes laterally through the tip of the fibula. The lateral malleolus is posterior to the medial malleolus.

The integrity of the ankle mortice is maintained by the interosseous

253

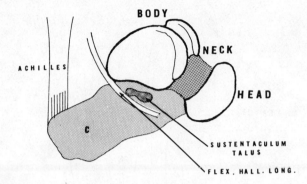

FIGURE 212. Talus. Comprised of a body, neck, and head, the talus body has two articulating surfaces that fit into the ankle mortice. The head articulates with the bones of the middle segment and the entire talus sits upon the calcaneus (C).

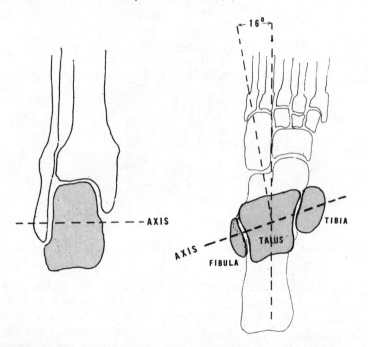

FIGURE 213. Ankle mortice and talus relationship. *Left,* The axis of rotation passes through the fibula and below the tip of the tibia. *Right,* Viewed from above the medial malleolus is anterior to the fibula forming a 16 degree toe-out stance and gait. The talus also is broader anteriorly than posteriorly.

ligament and membrane and the anteroposterior tibiofibular ligaments. The interosseous ligament attaches from the inner border of the tibia and proceeds laterally and downward to the inner aspect of the fibula. The fibula is loosely connected to the tibia and ascends during ankle dorsiflexion as the wider aspect of the talus separates the two malleoli. With full ankle dorsiflexion, causing full separation of the mortice, the interosseous ligament fibers are horizontal (Fig. 214).

Plantar flexion of the ankle, presenting the narrower portion of the talus, is accompanied by reapproachment of the tibia and fibula with the interosseous fibers resuming their oblique direction.

The ankle joint receives strong support from the collateral ligaments. The lateral collateral ligament is composed of three bands: (1) anterior talofibular ligament, which attaches from the tip of the

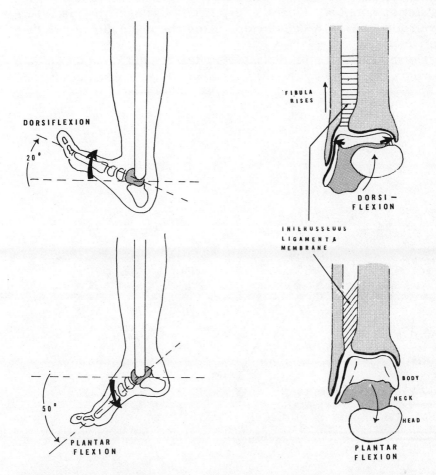

FIGURE 214. Motion of talus within mortice: dorsiflexion and plantar flexion. As the ankle dorsiflexes, the wider anterior portion of the talus wedges in the mortice and the fibula rises causing the interosseous ligament to become horizontal. When the mortice has been fully expanded, it prevents further dorsiflexion. On plantar flexion the narrow portion of the talus presents itself and the fibula descends. Here the fibers resume their oblique direction and the mortice decreases its width.

fibula to the neck of the talus, (2) calcaneofibular ligament, which runs from the tip of the fibula to the calcaneus, and (3) the posterior talofibular ligament, which connects the fibula to the tip of the talus (Fig. 215).

Severe ankle sprains sustained with the foot plantar flexed, and thus with lateral medial instability, usually injure the talofibular and calcaneofibular ligaments.

The medial aspect of the ankle joint is strongly supported by the deltoid ligament. This ligament is composed of four bands: (1) tibionavicular, (2) anterior talotibial, (3) calcaneotibial, and (4) the posterior talotibial. This is a very strong ligament that, in severe eversion injuries, usually avulses from the malleolus rather than sustain a tear.

Dorsal (20 degrees) and plantar (50 degrees) flexion of the ankle is partially restricted by the medial collateral ligaments. Because this

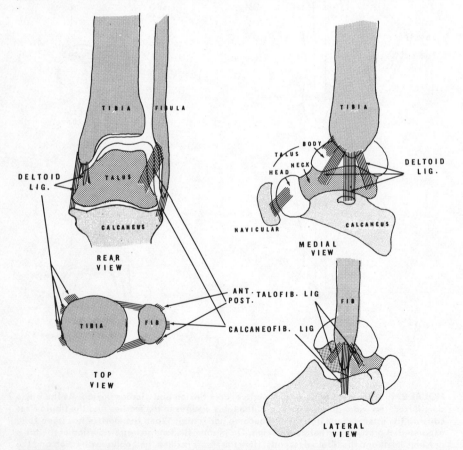

FIGURE 215. Collateral ligaments of the ankle joint.

ligament is eccentric to the ankle axis of rotation, the anterior fibers become taut to decrease plantar flexion and the posterior fibers limit dorsal flexion (Fig. 216).

Subtalar Joint

The talocalcaneal joint has three facets with different planes of articulation which permit only a small degree of inversion and eversion. Motion is also restricted by the ligaments.

The talocalcaneal joint is divided by a synovial-lined canal termed the tarsal canal. This canal is funnel shaped with the wide portion at the lateral end located just slightly below and anterior to the lateral malleous. The opening is increased by inverting the plantar flexed foot. It proceeds medially and posteriorly to its medial opening, located just behind and above the sustentaculum talus. The direction of the canal is approximately 45 degrees to the anteroposterior

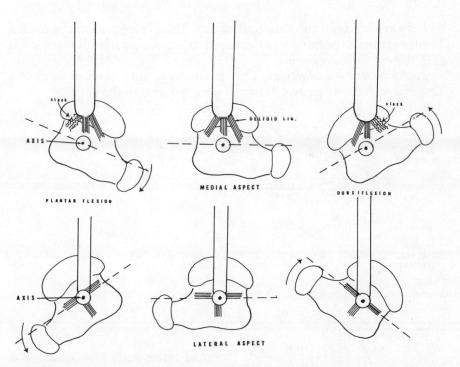

FIGURE 216. Relationship of medial and lateral collateral ligaments to the axis of ankle motion. The transverse axis of the ankle mortice transects the center of attachment of the lateral ligaments at the tip of the fibula. Plantar flexion or dorsiflexion of the ankle does not change the length of the ligaments. The axis is eccentric to the medial deltoid ligament. During plantar flexion the posterior strands become slack and the anterior fibers taut. The opposite occurs during dorsiflexion.

alignment of the foot. The canal is of sufficient width, thus making intra-articular injections relatively simple (Fig. 217).

A firm talocalcaneal ligament binds the two bones. This ligament runs the length of the tarsal canal but is more discrete at the fibular end to permit some rotation about this point. The talocalcaneal ligament by virtue of its direction within the tarsal canal becomes taut during inversion and slackens during eversion. By this action it makes the inverted (supinated) foot more stable.

Movement of the talocalcaneal joint can be isolated and tested by fully dorsiflexing the ankle, a position which locks the talus in the mortice. Then moving the calcaneus documents the extent of joint motion and localizes pain originating from the joint.

The axis about which the calcaneus rotates upon the talus forms a 45 degree angle to the floor and a 16 degree angulation in a medial direction from a line drawn through the second metatarsal (Fig. 218).

This is termed the subtalar axis about which three movements occur:

1. *Inversion* about the longitudinal axis. This is elevation of the medial border as the lateral border of the foot depresses. (*Eversion* about this axis is in the opposite rotation.)

2. *Abduction—adduction.* This is outward and inward rotation respectively about an axis drawn vertically through the tibia.

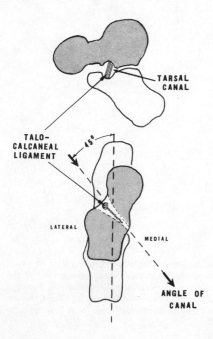

FIGURE 217. Subtalar joint—tarsal canal. The talus and calcaneus form a canal which has a 45 degree angle with the anteroposterior axis of the foot. The lateral opening is under the lateral malleolus and is readily palpable. A firm ligament bends the two bones.

258

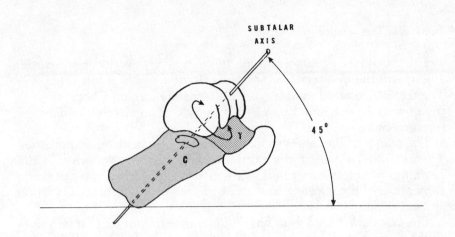

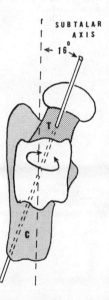

FIGURE 218. Subtalar axis. Movement about the subtalar axis consists of supination and pronation of the foot. The axis forms a 45 degree angle with the ground and a 16 degree angle medial to a longitudinal line drawn through the second metatarsal.

3. *Dorsal and plantar flexion* about the transverse axis. This last motion is significantly less than the similar motion of the talus within the ankle mortice.

Supination of the foot is a combined motion of the three subtalar motions, namely inversion, adduction, and plantar flexion. *Pronation* combines eversion, abduction, and dorsiflexion.

Transverse Tarsal Joint

The transverse tarsal joint combines essentially two joints in an axial alignment: the talonavicular and the calcaneocuboid joints (Fig. 219). The transverse tarsal joint has been alternately termed Chopart's joint, the midtarsal joint, and, because it is the site of an elective amputation, the surgeon's tarsal joint.

The planes of the talonavicular and the calcaneocuboid joints differ. The rounded head of the talus fitting into the concavity of the navicular bone permits rotation about this axis during pronation and supination of the forefoot and some gliding motion during inversion and eversion of the foot.

The calcaneocuboid joint has limited range of motion that permits a slight degree of abduction and adduction of the forefoot. When the axis of the talonavicular joint parallels the axis of the calcaneocuboid

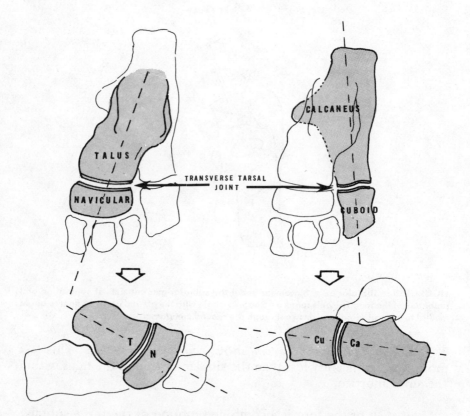

FIGURE 219. Transverse tarsal joint. The talonavicular and the calcaneocuboid joints combine to form the transverse tarsal joint. The broken lines depict the axis of rotation of each joint. These are parallel in the pronated foot and divergent in the supinated foot.

joint, as it does in the pronated foot, there ensues some motion that results in instability. In the supinated foot both axes diverge, decreasing joint mobility and thus creating a more stable foot. This factor plus the tautness of the talocalcaneal ligament explains the stability of the supinated foot. Also, in supination, all articular surfaces are "closepacked," that is, in complete opposition, and thus are engineeringly more stable and less dependent upon ligamentous support.

Middle Functional Segment

The middle functional segment of the foot (see Fig. 211) consists of five tarsal bones: the navicular, cuboid, and the three cuneiforms. These bones are firmly joined by their joint surfaces into a rigid transverse arch (Fig. 220). The mobility of the middle segment in its contact with the posterior functional segment produces the flexibility that permits the human foot to accommodate to uneven surfaces during walking activities.

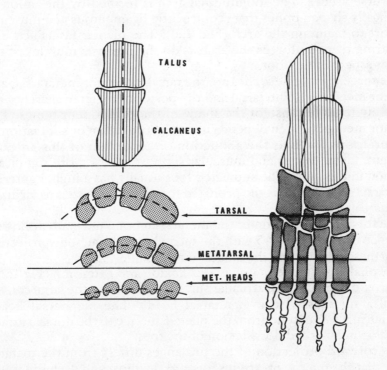

FIGURE 220. Transverse arches. *Left,* The fixed tarsal and posterior metatarsal arches and the flexible anterior metatarsal arch. *Right,* The receded second cuneiform that creates a mortice for the seating of the base of the second metatarsal.

261

The anterior margin of the middle segment is irregular in its union with the anterior segment. The second cuneiform is smaller and thus its anterior margin is set back from the other two cuneiforms. This creates a mortice into which fits the base of the second metatarsals.

The metatarsals articulate with the tarsal bones in a unique manner. The second metatarsal fits into the mortice between the first and third cuneiforms and thus can move in only one plane: dorsal or plantar flexion. The third, fourth, and fifth metatarsal bases are obliquely shaped to permit rotation upon the adjacent metatarsals. By virtue of this rotation potential the forefoot can "cup" and increase the transverse arch. The fixed second metatarsal remains the apex of the arch.

The first metatarsal also rotates about the base of the second metatarsal, but in the opposite direction to complete the arch. The base of the first metatarsal has a cartilaginous surface and is so shaped to permit dorsal and plantar flexion by gliding upon the first cuneiform about which it also rotates.

There are four arches to the foot: the longitudinal arch and three transverse arches. The longitudinal arch is formed by the union of specifically shaped bones that require little ligamentous and muscular support to maintain the arch (Fig. 221). The plantar fascia acts as a bowstring that reinforces the arch as do the intrinsic muscles of the plantar aspect of the foot.

The three transverse arches are the tarsal, posterior metatarsal, and anterior metatarsal. The tarsal and the posterior metatarsal arches are relatively fixed because of the shape of the component bones. The anterior metatarsal arch depends upon the pronation or supination of the forefoot as well as the abduction or adduction of this anterior segment. Ligamentous and muscular support have a more significant function in this arch. The supinated foot usually has a higher anterior metatarsal arch, whereas the pronated foot tends to flatten to virtually eliminate any arching.

The forward projection of the metatarsal bone normally follows a sequence of 2>3>1>4>5 with the second metatarsal bone projecting the furthest, albeit being the longest metatarsal.

The phalanges move in one plane: flexion and extension (Fig. 222). The proximal phalangeal articular surface glides upon the large convex articular surfaces of the metatarsal heads. The metatarsal heads, articular surfaces extend from the plantar surface to the dorsal surface, thus allowing marked dorsiflexion of the toes.

The forward projection of the phalanges differs from the metatarsals in that the big toe protrudes forward the most and each toe is less in sequence (Fig. 223).

If the physician has adequate knowledge of the bony anatomy of the

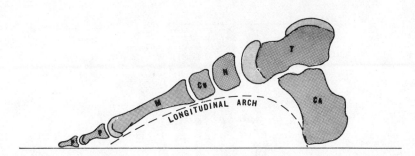

FIGURE 221.   Longitudinal arch. The longitudinal arch is formed by contiguity of the talus *(T)* with the calcaneus *(CA)* and anteriorly to the navicular *(N)*. Medially viewed the navicular articulates with the cuneiforms *(CU)*, thus with the metatarsals *(M)* and the phalanges *(P)*. The plantar fascia is not shown.

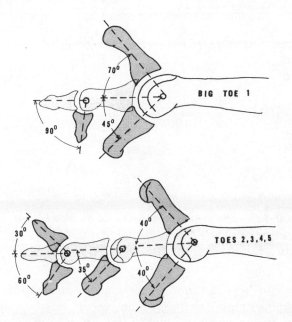

FIGURE 222.   Range of motion of phalanges. The phalanges move in only one plane, flexion and extension. *Above,* The big toe (hallux). *Below,* The remaining toes have three phalanges.

foot and ankle, he can examine the foot bimanually and test each joint separately and, thus, determine joint function and elicit the site of pain. The talocalcaneal joint can be tested by fully dorsiflexing the ankle, which immobilizes the talus, and by moving the calcaneus in a varus-valgus direction. The medial and lateral collateral ankle ligaments can be evaluated in a neutral or plantar flexed foot position.

With the hindfoot segment immobilized (heel held firmly in one hand)

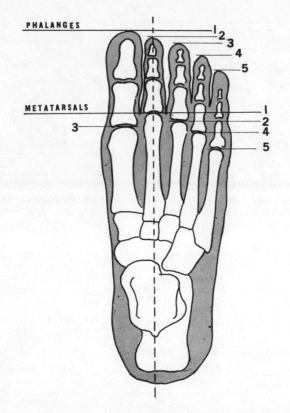

PHALANGES

METATARSALS

FIGUR 223. Relative length projection of the metatarsals and phalanges. The relative anterior protrusion of the metatarsals follows the pattern of 2>3>1>4>5. This makes the second metatarsal the longest and the first metatarsal the third longest. The toes have a pattern of 1>2>3>4>5 in which the first toe protrudes farthest forward followed by each adjacent toe in regular sequence.

the subtalar can be examined by pronation and supination of the forefoot. Keeping in mind that the second metatarsal joint moves only in a plantar dorsiflexion plane and that all the others have rotatory range as well, each metatarsotarsal joint can be tested. The phalanges can be tested individually by fixing the promixal phalanx and moving the more distal joint.

Muscles

The muscles that act upon the foot are the *extrinsic* muscles, which originate from the lower leg and attach to the foot, and the *intrinsic* muscles, which originate and insert within the foot.

The extrinsic muscles include the gastrocnemius soleus group, which plantar flexes the foot and inverts the forefoot. The lateral or peroneal muscles evert the foot; the medial or posterior tibial muscles invert as well as plantar flex; and the anterior group or tibial muscles dorsiflex and supinate the forefoot simultaneously.

The muscles act differently in the nonweight-bearing and weight-bearing foot. The origin and insertion alternate. In the nonweight-bearing position the origins are on the tibia and fibula and insertion and action is upon the foot. However, upon weight bearing, the origin is from the plantigrade foot and insertion and action is upon the leg.

The muscles that act upon the toes originate from the tibia, fibula, and interosseous membrane. The long extensor muscle inserts upon the distal phalanx and extends the big toe. The long extensor tendons of the toes divide into a central tendon, which inserts upon the middle phalanx, and two divided slips, which insert upon the distal phalanx. The long extensor muscle extends the toes and assists in everting the foot.

The flexors of the toes are both extrinsic and intrinsic. The long flexor muscle of the great toe originates from the tibia and fibula and its tendon passed under the medial malleolus to attach upon the distal phalanx of the big toe. It crosses two joints and acts to press the distal phalanx to the floor (Fig. 224).

The long flexor muscle of the great toe originates also from the posterior aspect of the tibia, passes behind the medial malleolus, and attaches to the distal phalanx of all other four toes, but, by passing three joints, acts to grip the floor.

The short portions of all the flexor muscles are intrinsic muscles originating within the plantar aspect of the foot.

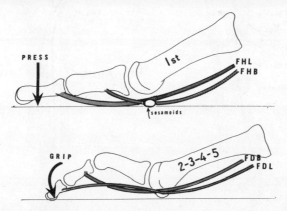

FIGURE 224. Action of the flexor tendons. The tendons of the big toe cross two joints and thus act by pressing the distal phalanx to the floor. The sesamoid bones are incorporated into the tendons of the short flexor muscle of the great toe (flexor hallucis brevis, FHB) and act to fulcrum the flexor action. The flexor tendons of the other four toes cross three joints and act to grip the floor. This action is performed by the short flexor muscles (flexor digitorum brevis, FDB) and long flexor muscles of the toes (flexor digitorum longus, FDL).

The nerve supply of the lower leg and foot are branches of the sciatic nerve (L₄ to L₅ and S₁, S₂, and S₃) and are depicted in Figure 225. The continuation of the tibial nerve proceeds to the intrinsic muscles of the , foot by way of the plantar nerve (Fig. 226).

EVALUATION OF THE PAINFUL FOOT

The normal foot must conform to the following criteria:

1. It must be painfree.
2. It must have normal muscle balance.
3. It must have no contractures.
4. The heel must be central.
5. The toes must be straight and mobile.
6. During gait and stance must have three sites of weight bearing.

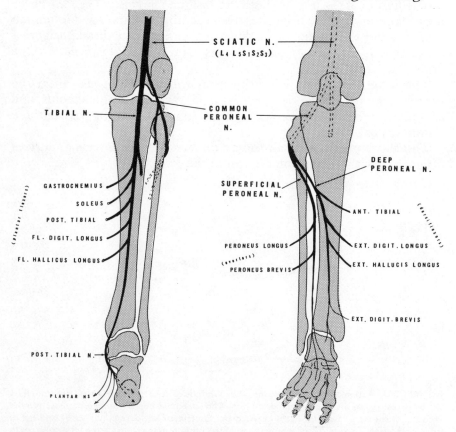

FIGURE 225. Innervation of the leg and foot. The sciatic nerve divides at the popliteal angle to form the tibial nerve and the common peroneal nerve.

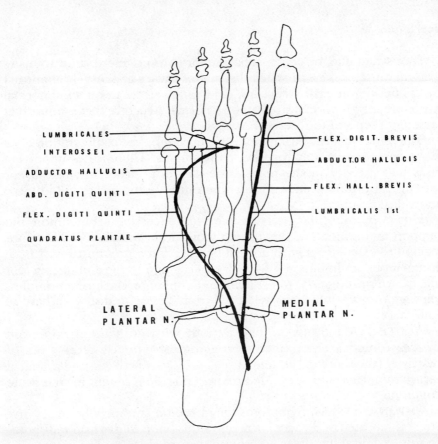

FIGURE 226. Muscles of foot innervated by the plantar nerves.

Pain, difficulty in walking, or an awkwardness of gait are the usual problems that cause the patient to seek consultation. Foot pain, when noted during standing, can be considered *static* and, when noted during walking, can be considered *kinetic*. The history and evaluation of the foot during relaxation, standing, and walking constitutes an adequate examination. Usually the patient can specifically indicate the site of pain which is anatomically discernible visually by palpation or passive movement.

Since this chapter is directed principally to pain the discussion of neurologic examination by performing motor function will be omitted. Reference is made to my book, *Foot and Ankle Pain.*[1]

The majority of painful conditions of the foot originate in the soft tissue: muscles, ligaments, tendons, nerves, blood vessels, and tissues of the joint spaces. In most cases of foot and ankle pain a local lesion can be implicated, and it can be the result of trauma or stress.

Foot strain may be acute, subacute, or chronic. Here, as in so many painful musculoskeletal states the rule of causes applies: (1) abnormal stress upon a normal structure, (2) normal stress upon an abnormal structure, or (3) normal stress upon a normal structure that is not at that moment prepared to receive the stress.

The static weight-bearing foot is supported by the configuration of its bony components held together by ligaments. There is no supporting muscular activity in the muscles of the foot, either intrinsically or extrinsically, during stance even when heavy objects are held by the individual.

However, muscular activity prevents excessive strain upon the supporting ligaments in the moving foot during ambulation. Therefore, pain in the static foot with normal bony architecture must result from ligamentous strain imposed by faulty mechanics. In the ambulating foot the muscles prevent the forces being imposed upon the ligaments unless they are overwhelmed, are weakened from disease or disuse, or have an imbalance.

ACUTE FOOT STRAIN. The patient with an acute strain from unaccustomed activity, such as jogging for exercise after years of inactivity (the weekend athlete), usually recovers with rest and a gradual return to normal activities. Medical help is seldom sought for this acute condition.

CHRONIC STRESS. Symptoms may become chronic when excessive stress is repeated or if there is a mechanical abnormality that predisposes to pain and disability from an otherwise normal stress. The mechanical effect on all structures begins with strain and ends with deformation.

The initial stress causes ligamentous inflammation with resultant pain. Persistent stress can cause ligamentous elongation and even some degeneration. Support to the joint is lost and the joint undergoes excessive motion or malalignment. The stress inflames the joint capsule, a condition which also causes pain. Persistence of joint irritation or malalignment causes structural damage to the articular surfaces, and degenerative arthritis results. Nature's attempt at minimizing irritation and its sequelae is to respond with bony overgrowth, further deforming the joint with arthrosis. This sequence interrupted early may be reversed but if allowed to proceed may lead to irreversible damage.

The initial symptoms are usually described as aching and are noted in the tendons or ligaments of the foot, the calf muscles, or occasionally in the anterior musculature of the leg. Deep tenderness of these inflamed tissues is palpable (Fig. 227).

The weight-bearing foot is a complex structure with all component parts interdependent (Fig. 228). The body weight is transmitted through

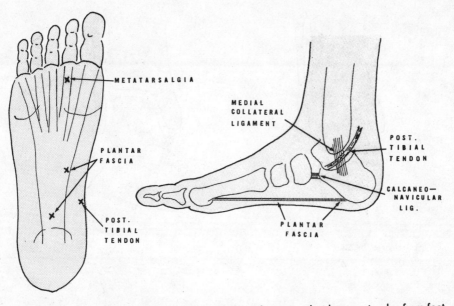

FIGURE 227. Tender areas in foot strain. All the soft tissues that become tender from foot strain are pointed out by patient and palpable by the examiner.

the tibia upon the talus which is in turn supported by the calcaneus. The calcaneus is oblique to the horizontal ground surface and thus predisposes the talus' gliding forward and medially. This force further everts the calcaneus and depresses its anterior portion. By virtue of these changes the plantar fascia becomes involved in supporting the longitudinal arch and becomes tender.

The increased obliquity of the calcaneus places stress on the medial longitudinal (deltoid) ligament, causing another site of pain. The forward gliding of the talus upon the calcaneus puts stress upon the inferior calcaneonavicular ligament, depressing the navicular with further decrease of the longitudinal arch.

As the calcaneus everts (valgus) the forefoot abducts, an action which decreases the two anterior transverse arches. The anterior metatarsal arch, when depressed, splays the forefoot and the arch disappears (Fig. 229). Now weight is borne on all metatarsal heads, although this is not their function. Pain results.

As the calcaneus goes into valgus the Achilles tendon undergoes adaptive shortening, thus causing further valgus and equinus and further strain on the anterior segments of the foot.

The mechanism and sequence of foot strain postulated above can occur in the normal foot which has become deconditioned from prolonged inactivity, protracted bedrest, or immobilization from casting

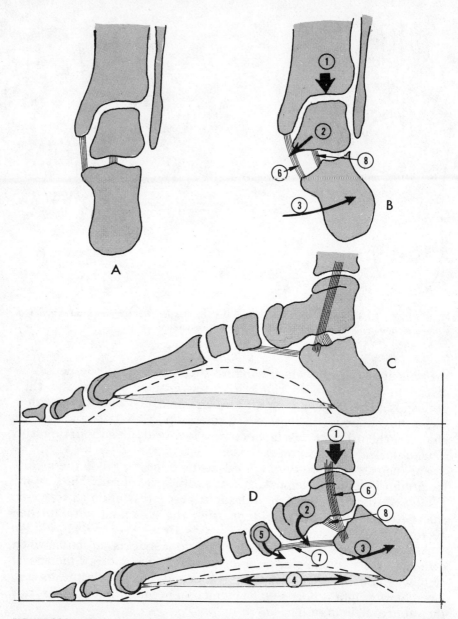

FIGURE 228. Mechanism of foot strain. A and C, Normal foot with proper bone and joint alignment, a central heel, and good longitudinal arch. B and D, Stress causes malalignment of structures. Weight-bearing impact of tibia (1) upon talus (2). The talus tends to slide forward and medially upon the calcaneus. Under pressure the calcaneus everts and rotates posteriorly (3), elongating the longitudinal arch and placing strain upon the plantar fascia (4). The rotating calcaneus depresses the navicular (5) by pulling upon the calcaneonavicular ligament (7) which becomes tender. The initial valgus of the heel places strain upon the medial collateral ligament (6) and ultimately upon the talocalcaneal ligament (8).

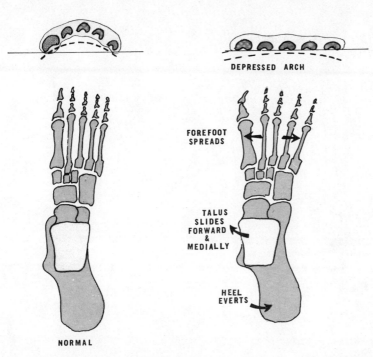

FIGURE 229. Pronation which causes splayfoot. As the heel everts and the talus slides forward and inward upon the calcaneus the forefoot abducts and broadens. The anterior metatarsal arch flattens and the inner three metatarsal heads become weight bearing.

for another unrelated orthopedic problem. Excessive weight gain or assuming a new profession requiring prolonged standing after years in sedentary profession may be contributing factors. Chronic foot strain, however, is more apt to result in the already pronated foot.

The pronated foot is supported by muscular activity. When this protective muscular action is overwhelmed the stress becomes imparted upon the ligaments, the joint capsules, and ultimately the joints themselves. Persistence of the stress converts the reversible functional deformity into a partially or completely irreversible structural deformity. Symptoms vary with the severity or the chronicity of the stress.

In the early phase of strain upon the pronated foot the muscles acting across the foot attempt to allay the strain upon the ligaments. As the foot assumes a more pronated posture the posterior tibial muscle, an invertor, acts to oppose further pronation. The direction and attachment of the posterior tibialis is depicted in Figure 230. Under stress the posterior tibial tendon becomes tender and can be palpated along its course under and behind the medial malleolus.

The anterior tibial muscle is an invertor as well as a dorsiflexor. It

271

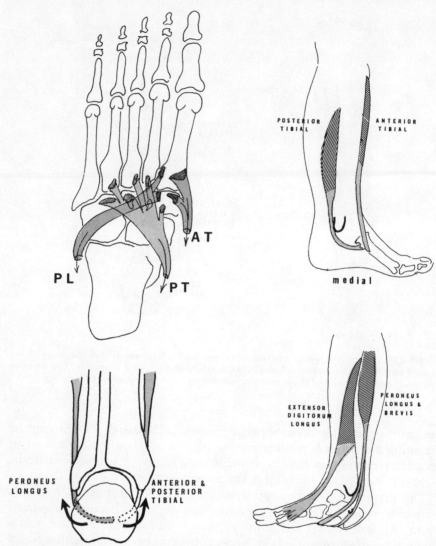

FIGURE 230. Extrinsic musculature of the foot. The origin, direction, and insertion of the extrinsic muscles acting upon the foot are shown. The anterior tibial *(AT)* and the posterior tibial *(PT)* are medial muscles that invert the foot. The long peroneal muscle (peroneus longus, *PL)* everts the foot. The posterior tibial and long peroneal muscles also plantar flex the foot. The extensor muscles of the toes and the anterior tibial muscle dorsiflex the foot.

functions in the latter manner during gait in the swing phase but has little function in the plantigrade weight-bearing foot.

As the foot goes further into pronation the lateral evertors shorten to take up the slack. The evertors are the peroneal muscle group but, with further forefoot pronation, the toe extensors change their alignment and become evertors of the foot as well as toe extensors. In prolonged stress the evertors may become inflamed and tender.

The talocalcaneal ligament is normally taut in the supinated foot and slack in the pronated foot. As the foot pronates the tarsal canal deforms. This subjects the talocalcaneal ligament to abnormal stress and it becomes inflamed. The tenderness here can be palpated by deep pressure into the lateral opening of the canal just anterior to and below the lateral malleolus. (See Fig. 217.)

The plantar fascia becomes elongated as the longitudinal arch flattens. Tender areas may be palpable.

The toe flexors also play a part in maintaining the longitudinal and transverse arches. Normally the long extensors extend the distal interphalangeal joints and the toe flexors can thus press the straightened toes against the floor. This action elevates the anterior metatarsal arch simultaneously. In the pronated foot the everted forefoot causes a malalignment of the toe extensors and they can hyperextend the metatarsophalangeal joint causing the flexors to claw the toe. The big toe extensor muscle has traction influence upon the longitudinal arch (see Fig. 249). In this position the metatarsal heads are more prominent and bear more weight (Fig. 231).

Ultimately the calcaneocuboid and talonavicular joints develop more "play" and sustain capsular and articular irritation with possible pain (Fig. 232). Pain originating from these joints can be verified by manually and forcefully everting (pronating) the forefoot while simultaneously immobilizing the heel. Tenderness can be elicited by pressure upon the plantar surface at the calcaneonavicular joint and its ligament.

TREATMENT. Foot strain treatment is essentially that of treating the pronated foot in the adult. In the acute strain local and general rest are indicated. This may include nonweight bearing, use of crutches, bedrest with elevation of the foot, and the application of heat or cold. Early in strains ice usually is of value in that it decreases congestion, decreases swelling, and has analgesic value. Heat later enhances healing. There are many advocates of alternating heat and cold.

The foot can be splinted to rest the joint(s) and ligaments by selective application of adhesive tape. Taping, so valuable in athletic injuries, is too often denied the non-athlete who sustains a severe sprain. Tape with ¼ to ½ inch adhesive-backed sponge rubber can be applied to elevate the portion of the foot desired. In severe sprains immobilization in a plaster cast, with or without a walking heel, is useful for several weeks.

Local injection of anesthetic agent with or without steroids is valuable providing the exact local anatomic structures are understood. Most ligaments and tendons are readily accessible to injection and the benefit derived is rewarding. Even the talocalcaneal ligament can be easily

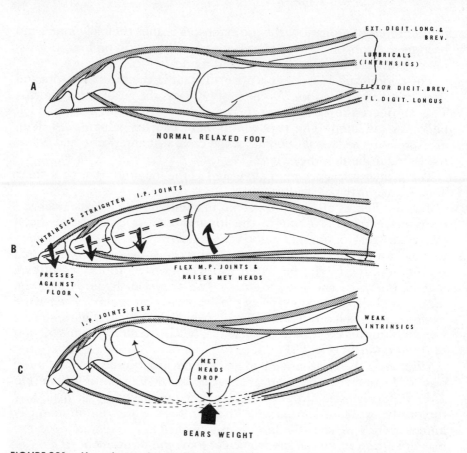

FIGURE 231. Muscular mechanism forming transverse arch. The transverse arch is only a potential arch, not an anatomic arch. The metatarsal heads are raised by the toes being kept straight and the metatarsophalangeal joints being flexed (B) by the long and short flexors. Weakness of the intrinsics permits the toes to bend at the interphalangeal joints and the flexors then increase the flexion at the interphalangeal joints. The metatarsal heads thus bear the full body weight (C).

injected by entering the canal at the lateral opening palpable directly under and just anterior to the lateral malleolus. Inversion of the forefoot opens the canal. The needle must proceed posteriorly and medially at 45 degrees. If the injection is given in this manner it is both safe and valuable.

Injection of joints is also feasible. Most joints of the posterior and middle segments are as accessible for intra-articular injections as are the joints of the anterior segments.

Changing shoes to proper footwear is valuable. The heel must be low and, in pronated feet, a Thomas heel inverts the heel and supinates the foot (Fig. 233). The Thomas heel also assists the gait in directing the foot

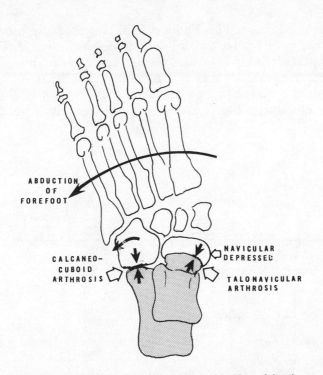

FIGURE 232. The chronically strained foot. The abducted forefoot of the chronically strained foot causes articular changes due to pressure of the abducted cuboid upon the calcaneus. The talus drops and causes pressure upon the superior portion of the navicular. Mechanical pressures at joints cause arthrosis.

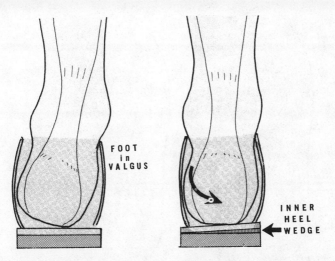

FIGURE 233. Inner heel wedge in treatment of heel valgus. The inner wedge should taper from no elevation at the outer border to $1/16$ to $3/16$ inches on the inner border. The exact elevation is determined by the height needed to place the calcaneus in a near-vertical position.

in a slight toe-in gait. The Thomas heel is preferably modified by elevating the inner border with a $^1/_8$ to $^3/_{16}$ inch inner wedge. This inner wedge may be extended to include the inner sole of the shoe.

Other inserts in the shoe such as are depicted in Figure 234 can alter pressure areas upon the foot or alter the weight-bearing or ambulatory foot. The metatarsal bar placed across the metatarsal heads avoids

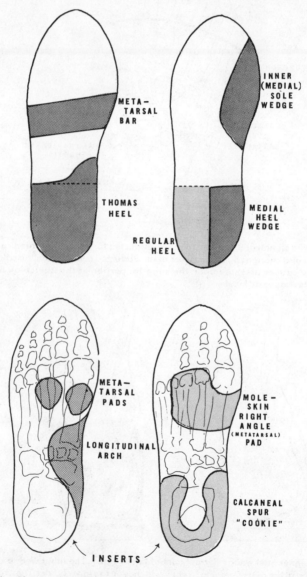

FIGURE 234.   Shoe modifications to correct strain factors.

276

pressure upon the heads and decreases the tendency for the toes to hyperflex at the interphalangeal joints and hyperextend at the tarsometatarsal joint.

Metatarsal pads placed behind the second, third, and fourth metatarsal heads elevate the forefoot and restore the transverse arch. The longitudinal pad insert restores the longitudinal arch and simultaneously inverts the foot to supination.

The shoe must be broad across the forefoot to allow the spread of the anterior segment of the foot. Since most shoe lasts have a toe-in counter they do not conform to the toe-out (abducted) forefoot of the pronated foot. Most shoes also have a tapered last with a pointed toe and thus constrict the forefoot (Fig. 235). Custom made shoes are often necessary when the foot is difficult to fit with a standard shoe.

Often when a sufficiently broad forefoot shoe is found the counter of the heel is too broad and the customer (patient) will sacrifice the comfortable forefoot width to acquire a snugger heel. To obviate this the proper width shoe can be modified by inserting thin felt pads on the lateral and medial aspect of the counter (Fig. 236).

Insofar as the weight-bearing foot is supported mostly by ligaments and joint capsules supporting the interlocking bones, little support is afforded by the muscles. The muscles, however, do exert some stress relief in the walking foot and therefore should be strengthened.

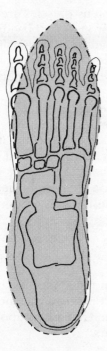

FIGURE 235. Foot constriction by standard shoe last. The pointed toe and narrow portion of the anterior portion of the shoe constricts the usual broad forefoot of the pronated foot.

277

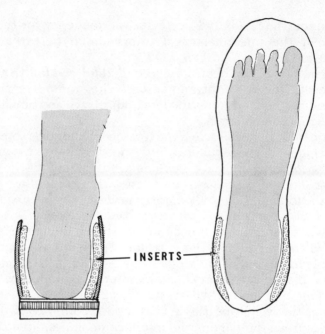

FIGURE 236. Counter insert. When a shoe is sufficiently broad for the splayed forefoot the counter is too broad to hold the heel. Inserting felt of ⅛ to ¼ inch width inside the shoe makes the heel snug and supported.

The posterior tibial and the gastrocnemius soleus muscles are foot invertors and are most beneficial in preventing pronation. Having the patient stand with feet slightly apart and slightly toed-in and rising up and down on the toes strengthens these muscles. Rolling both feet inward to place weight bearing upon the lateral border of the feet, and then flexing the toes, also strengthens the invertors.

Practicing gait by walking with a slight toe-in or exaggerated heel-toe is beneficial. A contracted heel cord aggravates pronation; therefore heel cords should be stretched. The Achilles tendon can be stretched in the following manner. The patient stands a distance from the wall and leans against it. The heel of the weight-bearing foot is held to the floor while the body leans forward. Rising up and down upon the weight-bearing foot while leaning forward gradually stretches the Achilles tendon and strengthens the gastrocnemius (Fig. 237).

Metatarsalgia

Metatarsalgia is a condition in which there is pain and tenderness of the plantar heads of the metatarsals. This usually occurs when the anterior transverse arch is depressed and causes excessive weight bearing upon the middle (second, third, and fourth) metatarsal heads.

278

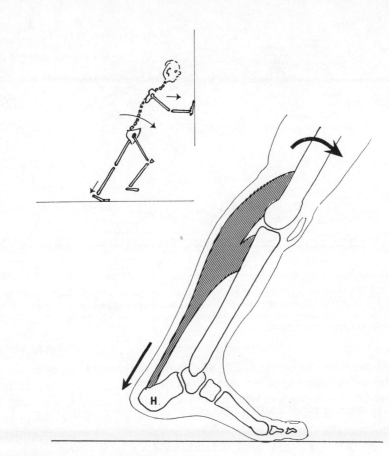

FIGURE 237. Exercise to stretch heel cord.

In the normal foot most weight is borne upon the first (big toe) and last (fifth) metatarsal which are padded. Five sixths of the weight is calculated as being borne normally by the first metatarsal head.

In the pronated foot the forefoot spreads and causes the interosseous ligaments to stretch and the transverse arch to depress (Fig. 238). Weight is borne upon the second and third heads which are not adequately padded for this function; therefore they become tender.

In the pronated foot with an everted forefoot the toe extensors become ankle dorsiflexors. They also extend the phalanges inadvertently upon the metatarsi and expose the metatarsal heads to a greater degree.

TREATMENT. Treatment of metatarsalgia is that of treating the pronated foot with emphasis upon elevating the middle of the anterior transverse arch. This arch is elevated by an inserted pad within the shoe placed under the second and third metatarsal bones *behind* the metatarsal heads (Fig. 239). If the pad is placed under the metatarsal heads it

279

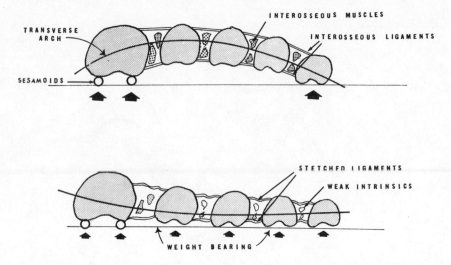

FIGURE 238. Splayfoot. A constitutional weakness of the intermetatarsal ligament combined with weakness of the intrinsic muscles of the foot may cause the foot to spread excessively under weight bearing. Symptoms consist of pressure pain of the middle metatarsal heads with formation of bunions and calluses.

aggravates the metatarsalgia. The site of placement of the pad is shown the patient by pressure of the examiner's thumb at the exact site desired. The patient then can feel where the pad should be and can modify its position within the shoe. A standard pad with adhesive to permit adherence to the shoe can be made into an inexpensive support in all the patient's shoes. A Thomas heel with an inner wedge is also indicated.

Morton's Syndrome (Fig. 240)

A short first metatarsal bone causes excessive weight to be borne by the second metatarsal. This is a congenital condition first described by Dudley Morton. The syndrome consists of (1) an excessively short first metatarsal that is hypermobile in its articulation with the cuneiform and the base of the second metatarsal, (2) posterior displacement of the sesamoids, and (3) thickening of the second metatarsal shaft.

Pain is usually felt at the base of the first two metatarsals and at the *head* of the second. Roentgen verification is diagnostic.

TREATMENT. Treatment requires an orthotic support under the first metatarsal bone to relieve weight bearing upon the second. Other aspects of treating the pronated foot also aid.

March Fracture (Fig. 241)

A march fracture is a stress fracture of a metatarsal shaft that often has minimal or no trauma elicited as its cause. Occasionally, pain is noted after

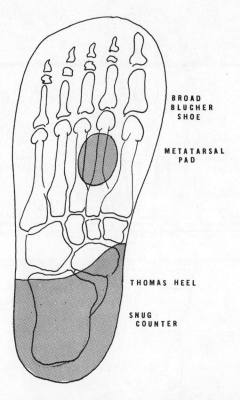

FIGURE 239. Shoe modifications in treatment of metatarsalgia. Placement of the metatarsal pad to elevate the heads of the second and third metatarsals must be *behind* the metatarsal heads. The broad forefoot shoe and a soft upper permits spreading of the foot without cramping, and the soft upper prevents calluses from forming on the dorsum of the toes. If a Thomas heel is used, the counter must be snug to hold the heel centrally.

a long march, hence its name. Initial x-ray pictures may be negative because the fracture is hairline with no displacement of fragments. Later as callus forms around the fracture roentgenography reveals that a fracture existed and caused the symptoms. Clinically there is tenderness at the middle of the involved metatarsal shaft and flexion or extension of the toes may be painful.

TREATMENT. Treatment consists of avoidance of weight bearing or walking, depending upon the severity of the symptoms. When symptoms are severe, a walking cast may be applied for three to four weeks.

Interdigital Neuritis

MORTON'S NEUROMA. The most common interdigital neuritis is Morton's neuroma (Fig. 242). This is a painful neuroma, a fusiform

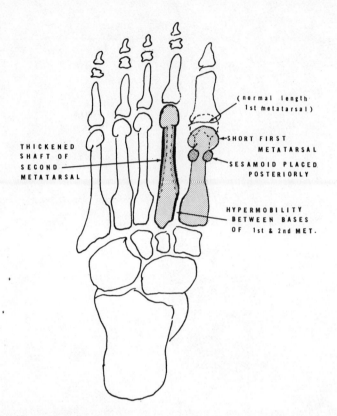

(normal length 1st metatarsal)

THICKENED
SHAFT OF
SECOND
METATARSAL

→SHORT FIRST
METATARSAL

SESAMOID PLACED
POSTERIORLY

HYPERMOBILITY
BETWEEN BASES
OF 1st & 2nd MET.

FIGURE 240. Morton's syndrome. A short first metatarsal bone causes excessive weight to be borne by the second metatarsal.

swelling of a digital nerve. The most common site is the nerve between the third and fourth toes, but it can occur between the second and third metatarsal also. The neuroma is usually located where the interdigital nerve branches (between the metatarsal heads) into branches to the two contiguous digits.

This condition is found most often in middle-aged women. A characteristic history is usually given of a "desire to take off one's shoes and massage the metatarsal area." Getting off one's feet does not afford relief as does removing the shoes.

On examination, pain is reproduced by pressure *between* the metatarsal heads (whereas in metatarsalgia tenderness is found by pressure on the plantar surface of the metatarsal heads). Numbness or hypalgesia may be elicited in the contiguous areas of the toes innervated by that interdigital nerve. Compressing the metatarsal heads together may elicit pain.

Merely prescribing adequately broad shoes to permit spread of the forefoot for treatment may suffice. A metatarsal pad behind the heads to

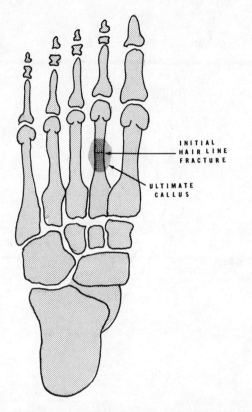

INITIAL
HAIR LINE
FRACTURE

ULTIMATE
CALLUS

FIGURE 241. March fracture of the second metatarsal. Often the initial fracture is not observed in routine roentgenography or appears as a hairline fracture. Within three weeks, after persistent pain, swelling, and tenderness, a callous formation indicative of healing becomes evident radiologically. This may be the first positive diagnostic sign. Displacement of fragments is rare.

elevate the transverse arch has some benefit. Diagnostically, and frequently therapeutically, an injection of an anesthetic agent with steroid gives relief. The metatarsal heads are easily palpated and the injection is given from the dorsum of the foot directly into the interdigital area. Persistence of symptoms warrants surgical excision.

OTHER INTERDIGITAL NEURITISES. *At the Transverse Metatarsal Ligament.* Another type of interdigital neuritis may occur from entrapment of the interdigital nerves as they pass the transverse metatarsal ligament (Fig. 243). This particular neuritis is caused by hyperextension of the toes. Clinically there is pain and tenderness from pressure *between* the metatarsal heads and by hyperextending the toes. Treatment is essentially the evaluation of the patient's activities and advise in avoidance of any position in which the toes hyperextend.

Posterior Tibial Neuritis. The posterior tibial nerve may be entrapped in

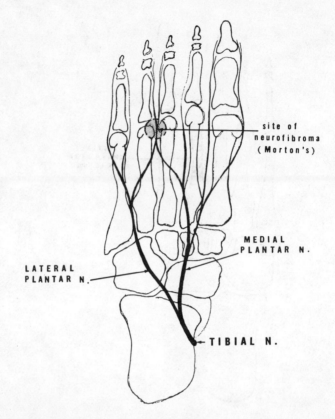

FIGURE 242. Morton's neuroma. This is a neurofibroma of the interdigital nerve. The most frequent site is the third branch of the medial plantar nerve as it merges with the lateral plantar nerve to form the digital nerve between the third and fourth toes. Pain occurs in the area and hypalgesia can be elicited in the opposing areas of the foot.

its passage behind and under the medial malleolus under the laciniate ligament (Fig. 244). This area comprising a bony depression behind the medial malleolus and covered by the laciniate ligament forms the tarsal tunnel and, besides the posterior tibial nerve, contains the tendons of the posterior tibial muscle and the long flexor muscles of the toes as well as the tibial artery and veins. The posterior tibial nerve carries the sensation to the sole of the foot and innervates the intrinsic muscles of the foot.

The nerve can be injured, compressed, or entrapped from a direct fall upon one's feet, standing on a ladder rung with soft shoes or bare feet, acute pronation of a normal foot, or accentuation of an already pronated foot. Ill-fitting longitudinal arch supports may cause the pressure.

Diagnosis is suggested by eliciting a history of "burning" in the plantar distribution of the nerve. This burning can be felt during standing or sitting and is *not* related to weight bearing. If the calcaneal branch is

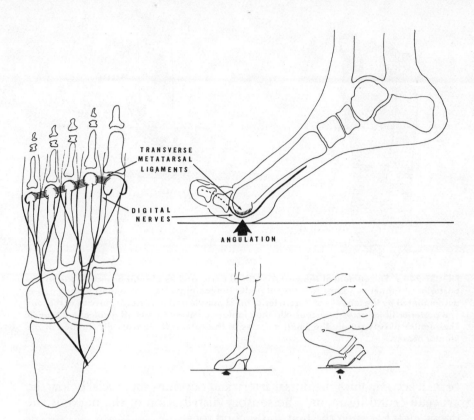

FIGURE 243. Entrapment of the interdigital nerve. The interdigital nerves are sensory to the toes. They come from the plantar nerves in the sole of the foot and pass across the transverse metatarsal ligament to the dorsum of the toes. If they are angulated at the ligament by posture and weight bearing, pain in the foot and numbness of the toes may result.

affected the condition may mimic that of a calcaneal spur. Tenderness is felt over the groove under the medial malleolus by pressure upon the nerve. Diagnosis can be verified by injecting an anesthetic agent into the tunnel; this should relieve the pain. Electromyographic conduction times are specifically diagnostic if prolonged. Sensory testing by touch or pin scratch may be used to outline the dermatome hypasthesia. Additionally, tapping of the tunnel may reveal a positive Tinel sign if the proximal flexors are weak.

Treatment by merely correcting the foot pronation through taping, casting, shoe correction, and modifying the gait may be successful. When tenosynovitis of the tunnel exists in conditions such as rheumatoid arthritis and the inflamed tendons compress the nerve, oral steroids or local injection of steroids are beneficial.

Anterior Tibial Neuritis. Anterior tibial neuritis may occur from injury to the distal branch of the deep peroneal nerve (Fig. 245). This nerve

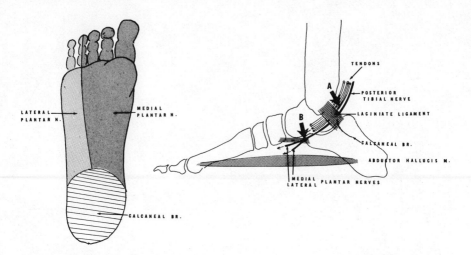

FIGURE 244. Entrapment of the posterior tibial nerve and its plantar branches. *Right,* The posterior tibial nerve passes in a tunnel under the laciniate (talocalcaneal) ligament. It is accompanied by the tendons of the posterior tibial muscle and long flexor muscles of the toes. The posterior tibial nerve branches into the plantar nerves and gives off a calcaneal branch. The plantar nerves supply the small muscles of the sole. *Left,* Sensory distribution of the plantar nerves.

branch accompanies the dorsal artery and becomes superficial below the cruciate crural ligament. The sensory distribution of the nerve is the dorsal cleft between the first and second toes.

The common injury is caused by direct pressure on the nerve, usually resulting from an ill-fitting shoe. Diagnosis and treatment may be affected by local injection of a Novocain derivative. Correction of the pressure prevents recurrence.

Referred Pain

Pain can be referred to the foot-ankle region from abnormal conditions in spinal nerve roots such as a herniated disk or spinal cord tumor. This referral pattern has been thoroughly discussed in Chapter 3 but it is mentioned here so that the physician will be alerted to that possibility when there is pain in the foot-ankle region.

The Painful Heel

Pain in the region of the heel may (1) arise from the tissues behind and under the calcaneus, (2) arise within the bones and joints of the heel, or (3) be referred to the heel region from a distant site (Fig. 246).

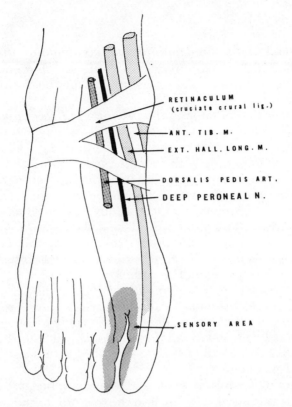

FIGURE 245. Trauma to the deep peroneal nerve. The deep peroneal nerve becomes superficial as it emerges below the cruciate crural ligament. There it is vulnerable to trauma causing pain and numbness in the area shown in the diagram.

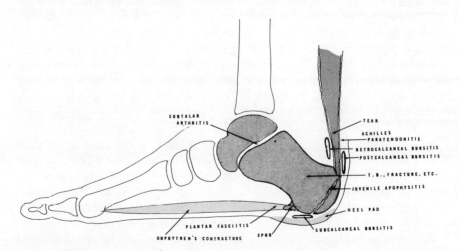

FIGURE 246. Sites of pain in the region of the heel.

Pain and tenderness noted *under the heel* is commonly attributed to a calcaneal spur which may or may not be evident with roentgenography. Frequently the condition is plantar fasciitis and may be the forerunner of a calcaneal spur.

Plantar fasciitis is common in the patient whose occupation requires long periods of standing or walking, especially when he or she has been unaccustomed to this activity and is deconditioned. The condition is more prevalent in people with pronated feet, a condition that places stress upon the longitudinal arch. Plantar fasciitis appears to be increasing with the advent of jogging for cardiopulmonary exercises, because the patient may not be well conditioned to jog in soft shoes and on hard surfaces. Males are more prone to acquiring plantar fasciitis.

This condition may be considered to be a tendofascioperiosteal irritation. The plantar fascia attaches by tendinous insertion to the periosteum of the calcaneus. The pathology is probably a minor tear or stretch of the plantar tendon fibers with avulsion of the periosteum from the bone. Subperiosteal inflammation occurs. In time it is repaired by fibrous tissue and then calcium deposit, ultimately forming a spur (Fig. 247). The initial pain and discomfort is probably due to the soft tissue inflammation of the fascial tendinous periosteal tissues and only in later stages is the pain due to the spur.

The presenting complaint is pain and tenderness beneath the anterior portion of the calcaneus radiating into the sole or into the calcaneal pad. The examination reveals deep tenderness of the anteromedial aspect of the calcaneus which is the site of attachment of the plantar fascia. X-ray pictures initially reveal nothing. In chronic recurrent episodes of acute fasciitis x-ray studies may remain negative, whereas spurs are noted in patients who are asymptomatic. Therefore the diagnosis is clinical.

TREATMENT. Treatment is primarily aimed at relieving local pain and secondarily at relieving the tension upon the plantar fascia or completing the tear. Injection of an anesthetic agent and steroid into the painful area is diagnostic as well as therapeutic. The injection may be given directly into the area through the heel pad, or the approach may be made from the medial or lateral aspect of the foot below the heel pad (Fig. 248). When given through the pad in a direct injection, the needle point is placed directly against the skin with moderate pressure for a few seconds; then increasing pressure will penetrate the pad with much less pain to the patient. The needle should penetrate *to* the bone and be moved anteriorly until it passes anterior to the calcaneus. At this point the needle is assumed to be in the site of plantar fascia insert.

Raising the heel of the shoe ¼ to ½ inch decreases the tension upon the calcaneus by removing tension upon the Achilles tendon from the

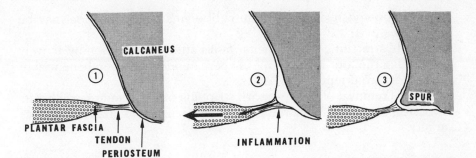

FIGURE 247. Mechanism of plantar fasciitis (heel spur). *1*, Normal relationship and attachment of the plantar fascia to the calcaneus; *2*, Traction upon the fascial tendinous portion to the periosteum separates the periosteum from the heel and resultant inflammation causes pain; *3*, Subperiosteal invasion by inflammatory tissue and ultimate calcification into a spur. This may be asymptomatic.

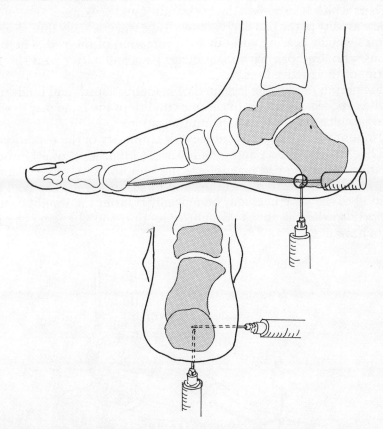

FIGURE 248. Injection technique in plantar fasciitis. The point of maximum tenderness of the plantar fascia can be injected directly through the heel pad. This is the area where the fascia inserts into the calcaneus. The site can be reached from the lateral or medial approach but localization by this approach is less accurate.

equinus foot position. A sponge rubber insert in the shoe may be beneficial.

Forceful stretching of the plantar fascia after the injection by actively dorsiflexing the foot with the toes hyperextended has been beneficial in the author's experience (Fig. 249).

Surgical removal of the spur with stripping of the plantar fascia has its advocates, but surprisingly there are many recurrences after this type of surgery.

Painful Heel Pad

Pain and tenderness may occur in the calcaneal pad. This pad is composed of fatty and fibroelastic tissue formed by fibrous septa into compartments. These compartments act as compressible shock absorbers which bear weight during standing and walking.

The elasticity of the pad decreases with age and gradually the calcaneus becomes weight bearing without the protection of the pad. There are persons who are born with inadequate pads and suffer pain the vast majority of their active lives.

Examination reveals the inflamed or inadequate pads and tenderness over the exposed calcaneus. In chronic conditions the bone may develop exostosis which aggravates the painful situation.

Treatment varies with the acuteness or chronicity of the condition. In acute heel pad irritation and inflammation, infiltration of a Novocain derivative under the pad may relieve the symptoms. A sponge-rubber heel pad may be inserted into the shoe to substitute for the heel pad. Elevation of the shoe heel may beneficially transfer the weight bearing anteriorly to the calcaneus. Weight reduction and changing the gait pattern has value.

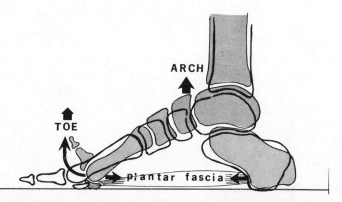

FIGURE 249. Effect of the toes upon the longitudinal arch. Full extension of the big toe exerts traction upon the plantar fascia which causes elevation of the longitudinal arch.

ACHILLES TENDONITIS. The Achilles tendon may become tender. Since the Achilles tendon does not have a synovial sheath this cannot be considered to be tenosynovitis. Inflammation, usually from trauma or stress, occurs in the loose connective tissue about the tendon, known as paratenon.

The tendon is tender to squeezing by the examiner and may appear to be thickened or swollen. Pain is aggravated by acute stretching from forcefully dorsiflexing the ankle. Running, jumping, and dancing aggravates the symptoms.

A retrocalcaneal bursitis may be present and cause similar symptoms. This bursa exists between the Achilles tendon and the skin. Irritation of this bursa usually occurs from ill-fitting shoes in which the shoe counter caused pressure and irritation.

Treatment of the tendonitis consists of a short-leg walking cast for four weeks to immobilize the foot and ankle. When there is visible distension in the bursa and fluid is present, aspiration followed by injection of steroids is beneficial. Correction of the shoe irritation is indicated; this may require cutting out the offending portion of the shoe. Moleskin tape may be placed on the Achilles tendon at the site of chronic irritation; this prevents further irritation.

RUPTURE OF THE ACHILLES TENDON. Rupture of the Achilles tendon may occur from stress or injury. It may occur from (1) direct trauma to the tendon such as a direct kick in an athletic injury, (2) abrupt stretch to an already fully stretched Achilles tendon, or (3) forceful ankle dorsiflexion when the ankle is relaxed and the patient unprepared for the stress. Achilles tendon tears occur most often in men between the ages of 40 to 50, especially when strenuous activities are undertaken after years of sedentary living.

Tearing occurs usually in the narrowest portion of the tendon approximately 2 inches above its point of attachment (Fig. 250). The tear may be partial or complete. Most partial tears ultimately become complete.

The person who sustains an Achilles tear experiences acute agonizing pain felt in the lower calf region, which immediately renders walking impossible. The patient cannot rise up on his toes if the tear is complete.

Examination is best done with the patient kneeling on the examining table with feet hanging over the edge. A gap in the tendon can frequently be palpated and the gastrocnemius soleus muscle is retracted. Ecchymosis may be found around the heel. If the tear is complete, the ankle can be dorsiflexed to a higher degree than the normal side. Squeezing the normal Achilles tendon causes reflex ankle plantar flexion, but in a complete tear this reflex action does not occur. This reaction has been termed the Simmond's test.

Partial tears, which are painful initially, become painfree when

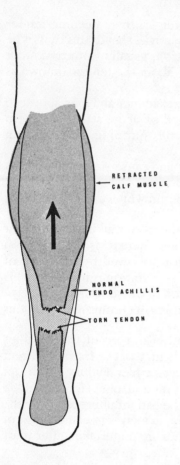

RETRACTED
CALF MUSCLE

NORMAL
TENDO ACHILLIS

TORN TENDON

FIGURE 250. Torn Achilles tendon. Most Achilles tendon tears are complete and occur approximately 2 inches above the calcaneal insert. The calf muscle retracts towards the popliteal space and a gap often can be felt at the site of the tear. Patient cannot rise on his toes.

completed. Inability to walk or rise up on toes may be the only residual. In immediate complete tears the pain may be momentary and the patient merely experiences a sudden fall to the ground as his calf muscle gives out.

Treatment. Treatment of the torn Achilles tendon has varied in recent years. Surgical repair followed by casting has met with varying benefit and is still considered by many orthopedists to be the preferable approach in the young person who intends pursuing an athletic career.

A nonsurgical approach of casting the foot for 8 weeks in the position of plantar flexion has led to satisfactory reunion of the Achilles tendon with good functional result. *Forced* equinus position is avoided and no less than 8 weeks of casting is effective. After the case is removed a 2.5 cm. heel lift is worn and the patient is cautioned about the possibility of falling. Crutches for a few days are beneficial.

Active gastrocnemius-strengthening exercises are instituted. The healed tendon remains thicker and the patient cannot rise as high on his toes as the normal side, but normal activities can be resumed. This form of treatment is based on the fact that the Achilles tendon regenerates itself and reunites.

PLANTARIS TEAR. A condition with acute pain in the calf experienced during physical activity that may mimic a tendon tear is the tearing of the plantar muscle or its tendon.

The plantar muscle originates from the posterior femur area just above the lateral epicondyle. The muscle is only 4 to 6 cm. in length and becomes a long tendon that descends the entire leg to attach to the calcaneus. In it course it lies between the gastrocnemius and soleus muscles.

The tear occurs during strenuous activity and pain may be agonizing though brief. Ecchymosis may result in the calf area or migrate to the Achilles area between the tendon and the tibia. The patient can rise up on his toes although pain at first may interfere and suggest a gastrocnemius-soleus tear. Passive ankle dorsiflexion is not increased as in a tear but may be restricted because of spasm. Deep tenderness of the calf may occur.

Treatment consists of wrapping with an ace bandage and restriction of activities as dictated by pain. Residual disability does not occur.

Fractures of the calcaneus must always be suspected in a painful severe injury and can be verified by roentgenography. Detailed discussion of calcaneal fractures are beyond the scope of this book.

PAINFUL ABNORMALITIES OF THE TOES

Hallux Valgus

Hallux valgus, frequently grouped in the painful condition termed bunions, is the most common painful deformity of the big toe. In essence hallux valgus is lateral deviation of the proximal phalanx upon the first metatarsal (Figs. 251 and 252).

Bunions, or hallux valgus, comprise a complex of (1) angulation of the first toe laterally towards the second toe, (2) enlargement of the medial portion of the first metatarsal head, and (3) inflammation of the bursa over the medial aspect of the metatarsophalangeal joint

This symptomatic condition is found mostly in older women who have had pronated feet with broadened forefoot and depressed metatarsal arch. Bunions are attributed in part to wearing shoes with narrow pointed toes that constrict the forefoot and high heels that insure the foot's being forced into the shoe.

The basic hallux valgus is considered to be a congenital condition. Metatarsal primus varus (medial deviation of the first metatarsal) in childhood may predispose to later hallux valgus. Abnormality of the

293

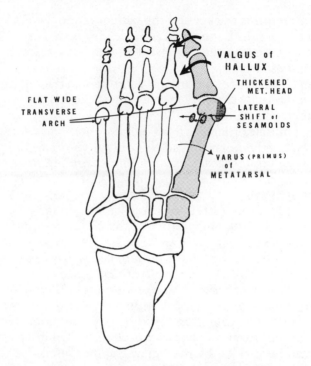

FLAT WIDE
TRANSVERSE
ARCH

VALGUS of
HALLUX

THICKENED
MET. HEAD

LATERAL
SHIFT of
SESAMOIDS

VARUS (PRIMUS)
of
METATARSAL

FIGURE 251. Major bony and articular changes characterizing hallux valgus. Hallux valgus is essentially a subluxation of the two phalanges of the big toe in a valgus direction. The first metatarsal deviates in a varus direction and the sesamoids are thus shifted laterally.

convexity of the first metatarsal head or muscle imbalance have also been implicated.

TREATMENT. Treatment of hallux valgus must be individualized and depends on the age of the patient, the severity and duration of symptoms, and the degree of deformity. Many patients with severe deformity are asymptomatic except for the difficulty of finding comfortable or attractive footwear.

Nonsurgical treatment consists of utilizing all the treatment components advocated for the pronated foot with attention especially to a wide shoe on the forefoot. If a painful bunion exists a pouch can be pressed out or cut out in the shoe at the site of the bunion. Molded shoes, although expensive and unsightly, may be necessary for comfort. Splinting of the toe in juvenile hallux valgus has its advocates but has had little success.

Surgery and its indications and numerous techniques are well documented in the literature and will not be discussed in this book. It must be noted, however, that surgery may relieve the symptoms but not necessarily alter the need for appropriate footwear after the operation.

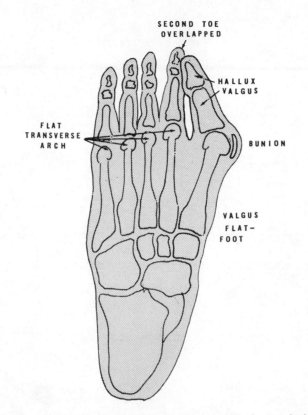

FIGURE 252. The type of foot with hallux valgus and bunion. The foot in which hallux valgus predominates is a broad forefoot with a depressed transverse metatarsal arch and a flattened longitudinal arch. The valgus big toe overrides or underlays the second toe which may be secondarily a hammer toe. A swollen inflamed bursa may overlie the enlarged head of the first metatarsal.

Hallux Rigidus

In normal gait every step causes the big toe to hyperextend during the last phase of stance as the body moves ahead of the center of gravity. Walking is impaired if the big toe metatarsophalangeal joint is rigid or if walking becomes painfully limited (Fig. 253). However, only the partially rigid toe is painful; the completely fused joint becomes painfree.

The limited extension of the big toe causes other foot problems as the patient attempts to avoid stressing that joint by walking on the outer border of the foot with increased toe-in gait and thus places more stress upon the fifth metatarsal. This gait is tiring as well as causing joint and callous discomfort in the outer aspect of the forefoot.

TREATMENT. Treatment of the painful hallux rigidus requires constructing a pad under the first metatarsal to prevent dorsiflexion

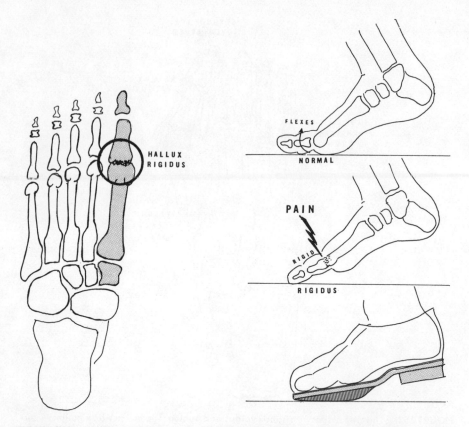

FIGURE 253. Hallux rigidus. As a result of damage of the metatarsophalangeal joint, which becomes rigid, the toe will not flex on toe-off of the gait, and pain can occur at each step. *Below,* Treatment consists of preventing stress on the rigid toe by placing a steel plate in the shoe sole to prevent bending and a rocker bar to permit painfree gait.

during walking. A steel shank in the sole of the shoe prevents bending the shoe last and a rocker sole added to the sole permits the foot to roll over the rocker during gait without extending the big toe (see Fig. 253).

Surgical intervention varies from resection of the joint and remodeling the metatarsal head to replacement of the joint with an orthosis.

Hammer Toe

The hammer toe is a fixed flexion deformity of the interphalangeal joint with hyperextension of the metatarsophalangeal joint (Fig. 254). This condition may be congenital or acquired. A painful callus may form on the dorsum of the middle interphalangeal joints or the tip of the distal phalanx.

296

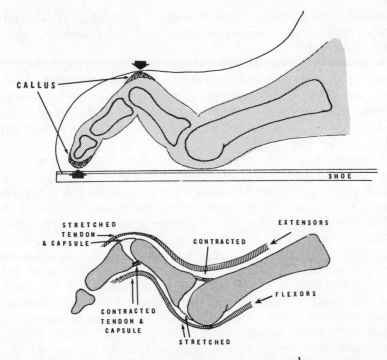

FIGURE 254. Hammer toe. The hammer toe is most often a flexion deformity of the interphalangeal joint with the capsule and tendons contracted on the concave surface. Subluxation is frequent. The proximal phalanx is usually extended and the distal phalanx flexed and flexible. Pressure and friction result in formation of painful calluses.

TREATMENT. Treatment may be required in order to eliminate pain or even ulceration at the callus sites or may be desired to permit wearing normal foot gear. Treatment to avoid pressure over the protruding joints is obtained by creating a bulge in the shoe at the site, cutting out the offending portions of the shoe, or prescribing molded shoes.

Physical therapy to manually stretch the toes and exercise the imbalanced muscles is usually futile. Surgical intervention is effective and should be sought in severe or disabling hammer toes.

ANKLE INJURIES

The most common painful injury to the ankle is sprain. This injury may vary from a simple strain in which the ligaments are merely elongated with minimal microtrauma to tearing of the ligamentous fibers with or without avulsion of the bones to which they attach. The most severe injury is fracture dislocation of the ankle. The severity of a sprain is so frequently unrecognized that the statement of Watson-Jones must

297

be heeded: "It is worse to sprain an ankle than to break it." This implies that fracture receives adequate treatment and sprain receives neglect or inappropriate care.

The lateral and medial collateral ligaments stabilize the ankle while permitting dorsiflexion and plantar flexion. The talus is firmly seated within the ankle mortice in full dorsiflexion but, as it presents its narrower width in plantar flexion, is mobile in a mediolateral direction. The neutral ankle position and especially that of plantar flexion expose the ligaments to the possibility of strain and sprain.

All ligaments of the ankle are supplied with sensory nerves and thus can evoke pain and, when stretched, reflex muscle spasm. Sprain is defined as a "joint injury in which some of the fibers of a supporting ligament are ruptured but the continuity of the ligament remains intact."[2]

Most ankle sprains occur when the ankle is in an unstable position and the stress is borne by the ligaments. The ankle is most unstable when the foot is plantar flexed and both medial or lateral collateral ligaments are exposed to stress. With the foot plantigrade and fixed to the ground the superincumbent body weight with the leg rotating about the ankle can result in a sprain. This maneuver is exemplified by the football player whose cleated foot is fixed to the ground and as he runs he turns his ankle and sustains a severe sprain.

The most common ligamentous injury occurs to the lateral ligaments from inversion stress. If the foot is plantar flexed at the time of the stress the anterior talofibular ligament is affected whereas if the foot is in a neutral position the calcaneofibular ligament sustains the injury.

Various degrees of sprain may occur. The ligament may be overstretched (strained) without disruption of the integrity of the fibers. Usually this is minor and recovery is rapid and complete. Fibers may be torn, thus constituting a sprain; this is more severe and requires a longer healing period. Fiber tearing may vary from partial to a complete tear. In severe injuries the ligament may essentially remain intact but avulse a small fragment of the bone to which it was attached.

The simple strain does not impair joint stability whereas a sprain with fiber disruption in essence is a self-reduced subluxation with residual instability. All sprains must be considered to have ligamentous damage and thus should be so treated.

Eversion injuries, which impart the stress upon the medial collateral ligament, usually cause bony damage rather than spraining or tearing of the medial collateral ligament, which occurs in inversion injuries. This difference is due to the strength of the medial ligament.

Diagnosis is best made if the injury is witnessed or if the patient or a spectator has a clear recollection of the details of the incident. Hours or days after the incident memory for details is vague and the swelling and ecchymosis is less specific and less localized.

During the examination, tenderness over the involved ligament can be elicited. In a mild sprain the foot does not invert to a greater degree than the normal and no gap can be palpated between the foot and the malleolus. Protective spasm of the muscles may have occurred, however, that prevent excessive inversion of the foot and thus gives false security to the examiner.

A roentgenographic study should be made of every significant or questionable ankle sprain. The pictures should be taken, preferably with forced inversion of the involved ankle and compared to the forced inversion of the normal ankle. An abnormal tilt of the talus will be revealed in the pictures of a severely sprained ankle (Fig. 255).

Novocain can be injected into the tender area to overcome spasm before taking the pictures but usually gentle manual inversion can accomplish the same result. The finding of any degree of talar tilt as compared to the normal contralateral ankle is suggestive of ligamentous tear.

Severe injuries, especially to the medial ligaments can separate the ankle mortice and tear the tibiofibular ligament (Fig. 256). This tear widens the ankle mortice and leads to marked instability. Degenerative changes can result in later life from this injury.

Treatment

The immediate care of the sprained ankle involves prevention of swelling. Initial treatment usually advocated is elevation of the limb and

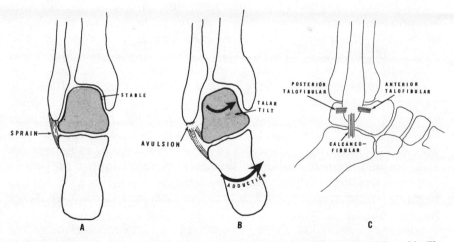

FIGURE 255. Lateral ligamentous sprain and avulsion. C, Lateral ligaments of the ankle. The anterior talofibular and calcaneofibular ligaments are the ligaments most frequently involved in inversion injuries. A, Simple sprain in which the ligaments remain intact and the talus remains stable within the mortice. B, Avulsion of the lateral ligaments where the talus becomes unstable and tilts within the morice when the calcaneus is adducted.

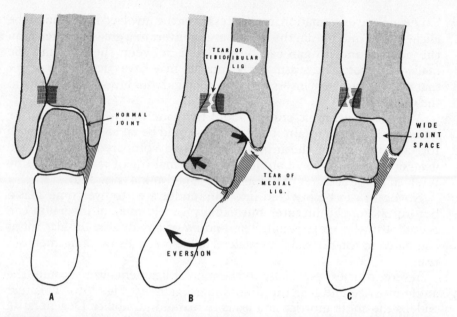

FIGURE 256. Avulsion of medial ligament and anterior tibiofibular ligaments from eversion injury to the ankle. *A,* Normal ankle mortice with the talus fit snugly between the malleoli. *B,* Eversion stress separates the malleoli *(arrows)* and tears the anterior tibiofibular ligaments and the medial deltoid ligament. Once the ankle has returned to its neutral position, *C,* a wide space remains between the talus and the medial malleolus which is a diagnostic sign with roentgenography. The ankle remains unstable with the wide mortice.

application of ice. Swelling occurs because of vascular injury with effusion and either microscopic or macroscopic hemorrhage. Occlusion of the arterial supply to the injured part appears to be physiologic. If swelling occurs effusion and hemorrhage will distend the joint, futher stretch the ligaments, and predispose to later formation of adhesions.

Curtailing blood supply to the ankle would appear feasible. Therefore, immediately upon an injury being sustained and observed, a sphygmomanometer could be applied to the leg immediately below the knee and compressed to pressure exceeding the arterial pressure. This occludes capillary circulation and prevents hemorrhage and edema. It also permits careful and more leisurely examination of the injury. Simultaneous to the application of the sphygmomanometer, the leg and foot can be elevated, placed in a firm bandage, and surrounded by ice packs.

After approximately 15 to 20 minutes of compression the sphygmomanometer can be briefly but completely released and then reapplied. This will prevent other undesirable symptoms of ischemia below the tourniquet. The period of occlusion allows early clotting of the injured capillary and arterial blood and diminishes the edema.

300

Elevation of the leg, immediate application of a firm yet evenly distributed bandage, and immediate application of ice are also advocated. If a sphygmomanometer is not available or there is no one present with knowledge of its use, elevation of the leg, ice packing, and pressure dressing are indicated.

A crepe or elastic bandage is preferable to adhesive tape because it can be applied and reapplied with minimal skin irritation. The bandaging must be firmly and uniformly applied to include the entire foot proximal to the tear and must include the lower half of the leg.

Ice can be applied within a plastic bag or preferably by emerging the entire foot in a container of water and ice cubes. If the leg is occluded by a sphygmomanometer, emergence in ice water is permitted. Ice should be applied at 20 to 30 minute intervals every 2 to 3 hours for the first day. Frequently, utilization of the cuff occlusion for a half to one hour makes further ice application unnecessary.

When walking is resumed, the ankle that has been severely sprained should be bandaged. Use of crutches to prevent or minimize weight bearing is often desirable.

Roentgenographic studies should be made as soon as possible. If effusion is marked in the presence of a fracture, plaster casting may be delayed. This delay in casting is acceptable if all aspects of immediate treatment are strictly observed.

If the sprain is minor, roentgenographic pictures are negative and excessive mobility of the ankle is not evident. The foot and ankle may be rebandaged daily and ice application continued for several days.

After three to four days, heat application may replace ice. The bandage should *not* be removed too soon as swelling may return or the ligaments resubjected to added stress. Seven to ten days of bandaging is usually adequate.

Active range of motion exercises performed by the patient helps disperse edema and prevent adhesions. This range of motion is for dorsiflexion, plantar flexion, toe flexion, and gentle inversion and eversion. Competitive sports should be delayed for one to three weeks depending on the severity of the sprain.

When a severe tear has occurred, the foot (in a lateral ligamentous tear) should be casted with the foot in slight *eversion* and at 90 degrees neutral dorsiflexion. If the cast is applied in the presence of edema it can be removed every day or few days and reapplied. Movement of the foot in the cast will indicate the frequency of recasting. In severe sprains the cast, with a walking heel, should be continued for approximately 8 to 10 weeks.

Frequently after a severe ankle sprain the patient complains of instability or a weak ankle that causes insecurity and clumsiness. Balance and coordination are impaired. When the patient is examined, there is poor balance on one leg standing as compared to the normal side. With or without the eyes being open, one foot balance is impaired. This indicates impairment of proprioception. As the ligaments are highly innervated, it is plausible that the sensory endorgans of the nerves within the ligaments are damaged by stretching during the ligamentous injury. Recovery is apparently incomplete.

Restoration or improvement of balance can be initiated by balancing exercises. Merely practicing one leg stance and balance will help. Standing on a board 4 × 12 × ½ inches, which is balanced on a half round, enhances the balance exercise (Fig. 257). At first the half round is placed at right angle to the length of the board. Balance in this position is one of ankle dorsiplantar movement. Gradually the half round is turned until it is longitudinal to the board (Fig. 258). At this point ankle eversion and inversion are stressed. Balance by this treatment can be gradually improved and the ligamentous proprioception enhanced.

Surgical repair of severely torn ligaments is usually not necessary if proper casting is applied. However, if instability remains after proper care

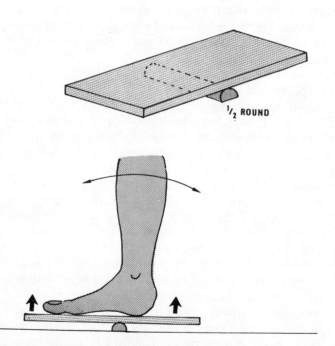

FIGURE 257. Balance exercise. Standing on a board 4 x 12 x ½ inches balanced on a half round placed at right angle to the board causes the patient to develop balance of ankle dorsiflexion and plantar flexion. Exercise can be performed for 5 minutes several times daily.

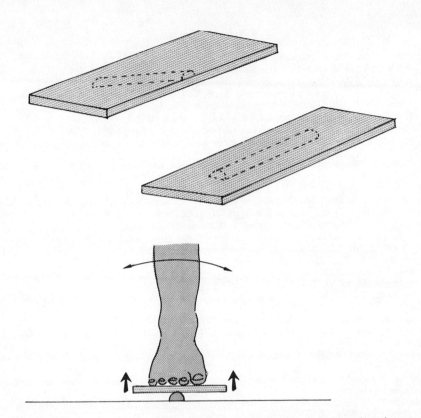

FIGURE 258. Balance exercise. Gradually the half round under the balance board is rotated until it is placed longitudinally to the board. This change in angulation of the half round requires more intricate balance of the ankle. The half round can be rotated at weekly intervals.

or recurrent dislocation occurs, surgical intervention should be considered.

In the injury that tears the medial ligament and the tibiofibular ligaments with widening of the ankle mortice the plaster cast should be applied with the intent of compressing the ankle bilaterally. Reapplication of casts here are indicated until the final cast is snug. Weight bearing even in the cast is to be avoided for a minimum of 8 weeks.

Open reduction of this injury may be necessary, but should only be considered if there is fragmentation or failure to reduce by casting.

REFERENCES

1. Cailliet, R.: Foot and Ankle Pain. F. A. Davis Co., Philadelphia, 1968, pp. 39–41.
2. Dorland's Illustrated Medical Dictionary, ed. 25. W. B. Saunders Co., Philadelphia, 1974.

BIBLIOGRAPHY

Basmajian, J. V.: Man's posture. Arch. Phys. Med. 46:26, 1965.

Basmajian, J. V., and Stecko, G.: The roles of muscles in arch support of the foot. J. Bone Joint Surg. 45[Am]:1184, 1963.

Brantingham, C. R., Egge, A. S., and Beekman, B. E.: The Effect of Artificially Varied Surface on Ambulatory Rehabilitation with Preliminary EMG Evaluation of Certain Muscles Involved. Presented at APA Annual Meeting, Los Angeles, August, 1963.

Chomeley, J. A.: Hallux valgus in adolescents. Proc. Roy. Soc. Med. 51:903, 1958.

Clayton, M. L., and Weir, G. J.: Experimental investigations of ligamentous healing. Am. J. Surg. 98:373, 1959.

De Palma, A. F.: Section I, Symposium: Injuries to the ankle joint. Clin. Orthop. Related Res. 42:2, 1965.

Du Vries, H. L.: Surgery of the Foot. C. V. Mosby Co., St. Louis, 1965.

Freeman, M. A. R., Dean, M. R. E., and Hanham, I. W. F.: The etiology and prevention of functional instability of the foot. J. Bone Joint Surg. 47[Br]:678, 1965.

Grant, J. C. B.: A Method of Anatomy, ed. 5. Williams & Wilkins Co., Baltimore, 1952.

Griffiths, J. C.: Tendon injuries around the ankle. J. Bone Joint Surg. 47[Br]:686, 1965.

Haymaker, W., and Woodhall, B.: Peripheral Nerve Injuries: Principles of Diagnosis, ed. 2. W. B. Saunders Co., Philadelphia, 1953.

Hicks, J. H.: Axis rotation at ankle joint. J. Anat. 86:1, 1952.

Hicks, J. H.: Mechanics of the foot. J. Anat. 87:345, 1953.

Hollinshead, W. H.: Functional Anatomy of the Limbs and Back. W. B. Saunders Co., Philadelphia, 1952.

Jones, F. W.: Talocalcaneal articulation. Lancet 2:241, 1944.

Jones, F. W.: Structures and Function as Seen in the Foot, ed. 2. Bailliere, Tindall and Cox, London, 1949.

Kaplan, E.: Some principles of anatomy and kinesiology in stabilizing operations of the foot. Clin. Orthop. 34:7, 1964.

Kelikian, H.: Hallux Valgus and Allied Deformities of the Forefoot and Metatarsalgia. W. B. Saunders Co., Philadelphia, 1965.

Keller, W. L.: Surgical treatment of bunions and hallux valgus. N. Y. Med. J. 80:741, 1904.

Lake, N. C.: The Foot, ed. 3. Bailliere, Tindall and Cox, London, 1943.

Lapidus, P. W.: Kinesiology and mechanical anatomy of the tarsal joints. Clin. Orthop. 30:20, 1963.

Lapidus, P. W.: Operation for correction of hammer toe. J. Bone Joint Surg. 21:4, 1939.

Lea, R. B., and Smith, L.: Non-surgical treatment of tendo achilles rupture. J. Bone Joint Surg. 54[Am]:1398–1407, 1972.

Levens, A. S., Inman, V. T., and Blosser, J. A.: Transverse rotation of the segments of the lower extremity in locomotion. J. Bone Joint Surg. 30[Am]:849, 1948.

Lewin, P.: The Foot and Ankle, ed. 4. Lea & Febiger, Philadelphia, 1959.

Liberson, W. T.: Biomechanics of gait: A method of study. Arch. Phys. Med. 46:37, 1965.

MacConaill, M. A.: The postural mechanism of the human foot. Proc. R. Ir. Acad. 1B:265, 1945.

Mann, R., and Inman, V. T.: Phasic activity of intrinsic muscles of the foot. J. Bone Joint Surg. 46[Am]:469, 1964.

McBride, E. D.: A conservative operation for bunions. J. Bone Joint Surg. 10:735, 1928.

McLaughton, H. L.: Trauma. W. B. Saunders Co., Philadelphia, 1959.

Mennell, J.: The Science and Art of Joint Manipulation, ed. 2. J. & A. Churchill Ltd., London, 1949.

Meyerding, H., and Shellito, J. G.: Dupuytren's contracture of the foot. J. Int. Coll. Surg. 11:596, 1948.

Milgram, J. E.: Office procedures for relief of the painful foot. J. Bone Joint Surg. 46[Am]:1095, 1964.

Morton, D. J.: Human Locomotion and Body Form: A Study of Gravity and Man. Williams & Wilkins, Baltimore, 1952.

Moseley, H. F.: Traumatic disorders of the ankle and foot. Clin. Symp. 17:3–30, 1965.

Murray, M. P., Drought, A. B., and Kory, R. C.: Walking patterns of normal men. J. Bone Joint Surg. 46[Am]:335, 1964.

Rubin, G. and Witten, M.: The talar tilt angle and the fibular collateral ligament. J. Bone Joint Surg. 42[Am]:311, 1960.

Ryder, C. T., and Crane, L.: Measuring femoral anteversion: The problem and a method. J. Bone Joint Surg. 35[Am]:321, 1953.

Saunders, J. B. de C. M., Inman, V. T., and Eberhart, H. D.: The major determinants in normal and pathological gait. J. Bone Joint Surg. 35[Am]:543, 1953.

Schwartz, R. P., and Heath, A. L.: Pointed and round-toed shoes. J. Bone Joint Surg. 48[Am]:2, 1966.

Sutherland, D. H.: An electromyographic study of the plantar flexors of the ankle in normal walking on the level. J. Bone Joint Surg. 48[Am]:66, 1966.

Taylor, R. G.: The treatment of claw toes by multiple transfers of flexor into extensor tendons. J. Bone Joint Surg. 33[Br]:539, 1951.

Wilson, J. N.: The treatment of deformities of the foot and toes. Br. J. Phys. Med. 17:73, 1954.

Wilson, J. N.: V-Y correction for varus deformity of the fifth toe. Br. J. Surg. 41:133, 1953.

Wright, D. G., Desai, S. M., and Henderson, W. H.: Action of the Subtalar and Ankle Joint Complex during the Stance Phase of Walking. Biomechanics Laboratory, Univ. of Calif., San Francisco, No. 38, June, 1962.

Zamosky, I. and Licht, S.: Shoes and their modifications. In Licht, S. (ed.): Orthotics Etcetera. Phys. Med. Library, Vol 9. Elizabeth Licht, Publisher, New Haven, 1966.

Index

307

308

309

Posture, 15
    acute neck trauma and, 131–133
    emotional stress and, 18
    erect, resuming, after acute low back
        pain, 89
    etiologic factors in, 131
    gravity and, 17
    improvement of, in treatment of low
        back pain, 70
    shoulder pain and, 161
    sitting, 16, 20, 21
    standing, 17
Psyche and neuromuscular pattern, 15
Psychologic response to pain, 101
Psychosomatic disorders, headaches and,
    139

QUADRICEPS femoris muscles, 224
Quervain's disease, 187–188
    treatment in, 188

RADIAL nerve
    hand pain and, 179
    tension, treatment of, 176
    trauma to, 174–176
Radiohumeral joint, 168
Radioulnar joint, 168
Raynaud's phenomenon, 194
Referred pain
    foot and ankle and, 286
    hip joint pain and, 213
    in acromioclavicular joint, 165, 166
    trigger points and, 32, 33
Residual pericapsulitis, subtle signs of,
    162
Resistive exercises in neurovascular com-
    pression syndromes, 148
Rest
    in acute low back pain, 81
    in hip joint pain, 208
    of hand, in treatment, 192
Reticular fibers, 5
Rheumatoid arthritis, 190–192
    knee and, 247
    treatment of, 192
Rhythmic stabilization, 160
Roentgenographic studies
    in acute trauma of neck, 121
    in ankle sprain, 299, 301
Rotation of the pelvis exercise, 70
Rotator cuff, 149–152
    tear, 162–164
    treatment in, 164

Round-back posture, 132
Rounded shoulder posture, neurovascular
    compression syndromes and, 148

SACRAL plexus, 206
Scalene triangle constriction, 144
Scitatic nerve, 206
    pain along, 77–81
Sciatica, corticosteroids in treatment of,
    97
Scoliosis, functional, in low back pain, 76
Sellar joint, 10
Shoulder
    functional anatomy of, 149–153
    girdle syndrome. See Neurovascular
        compression syndromes.
    pain in, 149–167
        active assisted motion in, 160
        exercise in, 155–157
        history in, 154
        immobilization in, 155
        injection in, 157–159
        medication in, 155
        physical examination in, 154
        physical therapy in, 155
        posture in, 161
        treatment in, 154–161
Simmond's test, 291
Sitting posture, 16, 20, 21. See also Posture.
Sit-up exercise for abdominal muscles, 72
Sliding displacement, 10
Smooth muscle, 8
Snuffbox, 186
Social Readjustment Rating Scale, 102
Soft tissue concept, 3–24
Specific theory of pain transmission, 26
Spheroid joints, 10
Spinning displacement, 10
Spinoreticulothalamaic transmission sys-
    tem, 26
Spinothalamocortic transmission system,
    26
Splinting
    in carpal tunnel syndrome, 184
    in rheumatoid arthritis, 192
Sprain
    ankle, 297–301
        examination of, 299
        treatment of, 299–303
    hand, 189
    treatment in, 36
Standing, 17. See also Posture.
    hip joint and, 201–204

311

Static foot pain, 267
Static low back pain, 67–75
Steroid injection in hip joint pain, 208
Steroid therapy in herniated disk, 98
Straight leg raising in acute low back pain, 76
Subclavian artery, pain in. *See* Neruovascular compression syndromes.
Subtalar joint, 257–259
Superior gluteal nerve, 206
Suprahumeral injection in shoulder pain, 157, 159
Suprascapular nerve block, 157
Surgical intervention, hip joint pain and, 212
Swan neck deformities, 192
Synovial membrane, hip joint pain and, 207

T CELLS and pain, 29
Talus, 252
Tarsal bones, 261–264
Tear(s). *See also* Rupture.
  meniscal, 239
  of long extensor tendon of thumb, 188
  periosteal, in tennis elbow, 176–178
Tendons, 6
  Achilles. *See* Achilles tendon.
Tennis elbow
  injection in, 176
  manipulation in, 177
  treatment of, 176–178
Tension headaches, 139
Thomas heel, 274
Thoracic vertebrae. *See* Low back pain.
Thumb, long extensor tendon of, tear of, 188
Tissue
  amorphous, 4
  areolar, 4
  composition of, 3
  concept of, 3–24
  connective. *See* Connective tissue.
  definition of, 3
  epithelial, 3
  fibrous, 4
  muscular, 3
  nervous, 3
Toe(s). *See also* Foot.
  hammer, 296
  painful abnormalities of, 293–297
Total hip replacement, 212

Tourniquet
  in knee injuries, 241
  in muscle contraction, 31
  in sprains, 36
Traction
  in acute low back pain, 84–86
  in acute trauma of neck, 128–131
  home, 129
  Buck's, 84
  gravity lumbar, 87
  hip joint pain and, 208
  pelvic, 84–86
    belt, 84
Translation, 10
Transverse arches of foot, 262
Transverse metatarsal ligament, neuritis at, 283
Transverse tarsal joint, 260
Trauma, 1
  definition of, 24
  elbow, 169–178
  neck, acute. *See* Neck, acute trauma of.
  radial nerve, 174–176
  trigger points and, 34
  ulnar nerve, 172–174
Triceps muscle, 168
Trigger fingers, 188
Trigger point(s), 32, 33
  headaches and, 137
  knee pain syndromes and, 233
  low back pain and, 81
  predisposing factors in, 35
Trochoid joints, 10
Tumors of hand, 194

ULNAR nerve
  compression, 185–186
    treatment in, 186
  hand pain and, 179
  palsy
    gradation of, 172
    treatment of, 173
  trauma to, 172–174
Uncovertebral joints, 110
  in degenerative joint disease, 133
Upper arm
  functional anatomy of, 107–120
  neck and, pain in, 107–141

VASCULAR impairment of hand, 194
Vascular tumors of hand, 194
Vasocoolant spray for trigger points, 35

Vertebral bodies, sensitivity of, 62
Vertebral column
   functional unit of, 41–54
   physiologic curves of, 54–56
Voluntary striated muscle, 8

WAHL-Melzak gate theory of pain trans-
   missions, 29
Walking. *See also* Gait.

center of gravity and, 21
   hip joint and, 198–201
Wall fingerclimbing exercise, 157
Wrist, hand and, pain in, 179–195

YELLOW fibers, 5
Yellow ligament, 49
   sensitivity of, 65